T0186231

PALAEOECOLOGY OF AFRICA

VOLUME 14

The large guelta of Molomhar, Adrar of Mauretania (20°35'N, 13°09'W), situated in the canyon Oum Lem Hâr, one of the major relict pockets of the western Sahara. Four species of fish, and four species of Amphibia occur here, beside several africotropical dragonfly species, and the freshwater medusa *Limnocnida tanganyiciae*. See article by H.J.Dumont. (Photo: H.J.Dumont)

PALAEOECOLOGY OF AFRICA
AND THE SURROUNDING ISLANDS

VOLUME 14

edited by

J.A.COETZEE (D.Sc.)
E.M.VAN ZINDEREN BAKKER SR (Phil.Nat.D., D.Sc.h.c.)
Institute for Environmental Sciences
University of the Orange Free State, Bloemfontein

A.A.BALKEMA/ROTTERDAM/1982

ISBN 90 6191 204 0

© 1982 A.A.Balkema, P.O.Box 1675, 3000 BR Rotterdam, Netherlands

Distributed in USA & Canada by: MBS, 99 Main Street, Salem, NH 03079, USA

Printed in the Netherlands

CONTENTS

V

FOREWORD

The valuable manuscripts received for volume 13 were too many to be incorporated into one issue of our sequence. This extra volume no. 14 contains part of them which are mostly concerned with northern and tropical Africa.

The glacial history of Africa is constantly being investigated from different angles and much more detail is reported in this volume on the glacial sequence of Mt Kenya and on the enormous influence cold and warmer stages had on the biogeography of the Sahara. Data provided from various fields on the transitional period round 24 000 BP give an insight into the impact of climatic change on the vast region of Africa north of the equator.

Some interesting papers dealing with palynology and archaeology are also included in this volume.

More reports have become available on the studies of late Quaternary lake level changes especially in the Ethiopian region. Attempts have been made to explain historic changes in these lake levels. Comparable research on the discharge of the Senegal River may make it possible to predict drought in the Sahel.

Holocene sea level changes along the West African coast are described using fossil faunas.

The extensive research programme carried out in the Turkana region is producing important information on the Plio-Pleistocene palaeoecology of that area.

The note on Ionium dating of peat is certainly of importance to all Quaternarists.

We hope that the added short notes and book reviews will be of value to our readers.

The contributors and the publisher deserve our thanks for their continued support.

Bloemfontein,
1 October 1981

The editors:
J.A.Coetzee
E.M.van Zinderen Bakker Sr

RELICT DISTRIBUTION PATTERNS OF AQUATIC ANIMALS: ANOTHER TOOL IN EVALUATING LATE PLEISTOCENE CLIMATE CHANGES IN THE SAHARA AND SAHEL

HENRI J. DUMONT

Zoological Institute, University of Gent, Belgium

SUMMARY

Although many kinds of adaptive strategies have evolved, it is generally true that higher aquatic animals can only disperse through water. They differ herein from plants and from the numerous terrestrial animals that are linked one way or the other to plants. A study of the relict distribution of selected groups of aquatic animals in a desert like the Sahara can therefore produce valuable new insights into the past climatic fluctuations of this area. Non-migrant dragonflies of African origin in the Mediterranean Basin are of two types: species with relicts on the central Mediterranean islands but not on the Iberian Peninsula and the coastal plain of NW Morocco, and species with relicts on the Iberian Peninsula, but not on the Mediterranean islands. It is argued that the first category corresponds to a migration wave around 20,000 BP, and the second to a much more important migration wave (embracing also many species of fish, amphibia, and the Nile crocodile), between 12,000-8,000 BP. Relictisation of these species in the Central Sahara occurred as early as 7,000 BP, while in the Southern Sahara and Sahel all relicts were wiped out by a severe drought around 5,000-4,000 BP. The Tibesti, Ennedi, and Adrar of Mauretania were recolonized by African species thanks to transgressions of Lake Chad and the River Senegal, but the Adrar-des-Iforhas and the Aïr still have a very deficient fauna, indicating that no surface water link with the Sudan has ever been re-established.

INTRODUCTION

While the phytogeography of the Sahara and Sahel (Quezel 1965, 1971) and the distribution in northern Africa of many terrestrial animal groups (mammals, birds, lepidoptera, coleoptera, terrestrial molluscs, and others) is known in considerable detail (see synthesis in Niethammer 1971), our knowledge of the aquatic biota has not significantly progressed since Gauthier's (1938) synthesis. Emphasis, at that time, was on deciding whether Saharan biota were

1

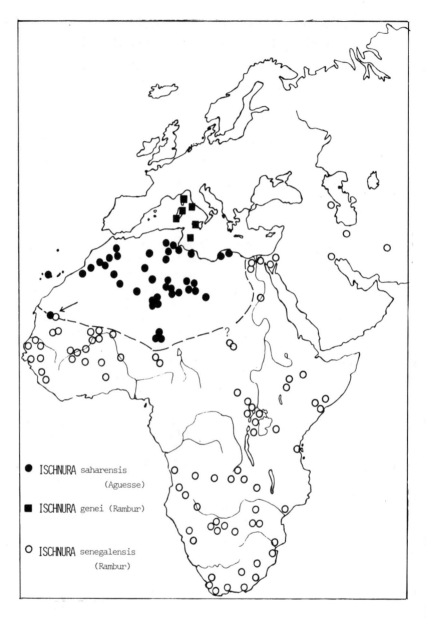

Figure 1. Range of three Ischnura species in Africa and on the Central Mediterranean islands.

2

predominantly Mediterranean or Africotropical in nature, and where the boundary between the Palaearctic and the Ethiopian regions should be situated (Monod 1938). Bodenheimer (1938) added the concept of an immixture of asiatic elements in the Saharan fauna to this discussion.

The dynamics of changing climates over the desert were suspected (Monod 1938), but their chronology and recurrence, even during the Pleistocene only, were beyond the imagination of workers three decades ago (see Butzer 1971 and various contributions in Van Zinderen Bakker & Coetzee 1980 and Williams & Faure 1980). Gauthier was therefore impressed — but had no explanation for — the absence of endemicity in the Saharan crustacean plankton and benthos, while this fact is now understood as a direct consequence of the short

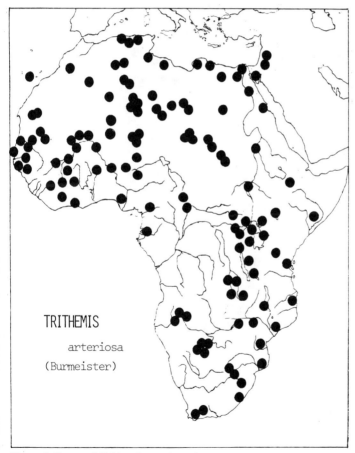

TRITHEMIS

arteriosa

(Burmeister)

Figure 2. Range of *Trithemis arteriosa.*

3

intervals at which climates changed, allowing insufficient environmental stability for much speciation to occur. In spite of this situation, speciation did take place in the calanoid copepods, and possibly in the ostracods as well, although this group deserves much more attention before any sound conclusions can be reached. A theoretical explanation for these exceptions is beyond the scope of the present paper and will be published elsewhere. Another exception concerns the phreaticolous fauna, which is largely endemic, although not strictly of the Sahara, but of its particular range within that desert. This very narrow endemicity is a general property of groundwater faunas all over the world, and a consequence of its relative independence of the climatic condi-

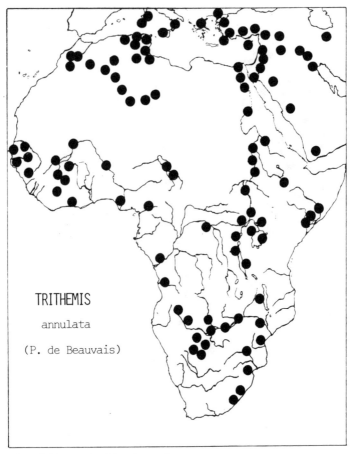

Figure 3. Range of *Trithemis annulata.*

4

tions at the surface. It may maintain its identity under all but the most drastic climate changes, and – if the rate of speciation is indeed some function of time – persist long enough to evolve endemics.

Gauthier's treatment further focusses on the Crustacea from temporary pools. Permanent aquatic biotopes are judged uninteresting and receive very little attention.

In the past six years, I endeavoured to produce a comprehensive and geographically representative study of the aquatic biota of the Sahara and Sahel regions. Eleven expeditions explored over 600 different aquatic sites, and information was collected on as many animal groups as possible.

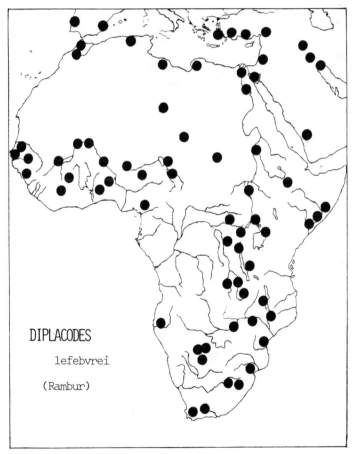

Figure 4. Range of *Diplacodes lefebvrei*.

5

I have given (Dumont 1981) an account of the zoogeographical patterns displayed by some copepoda and cladocera of the Sahara, and their significance to general biogeography and ecology.

The present account deals with one group of semi-aquatic insects (the Odonata) and with the aquatic vertebrates (amphibia, reptilia, and pisces).

MODES OF DISPERSAL

Aquatic animals have evolved a surprising variety of dispersing strategies. In fact, and although some general lines are linked to major taxonomical groups,

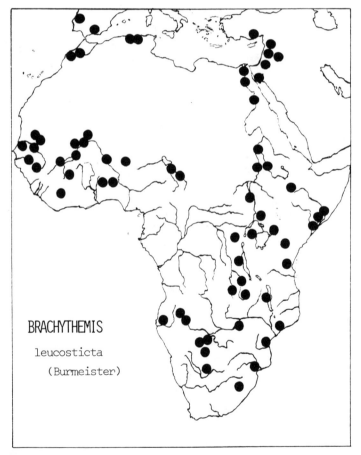

BRACHYTHEMIS

leucosticta

(Burmeister)

Figure 5. Range of *Brachythemis leucosticta.*

6

it is often necessary to consider individual cases (= species) and their adaptations to change in their environment. It is precisely from the combined knowledge of present relict distributions, mechanisms of dispersal, and their ecological implications that one can infer on climatic changes of the past. Conversely, if climate changes are known from other sources, one can assess the time of isolation of relict populations, and obtain insight into their rate of evolution.

I will, hereafter, deal first with the Odonata, then the Amphibia, reptiles, and fish.

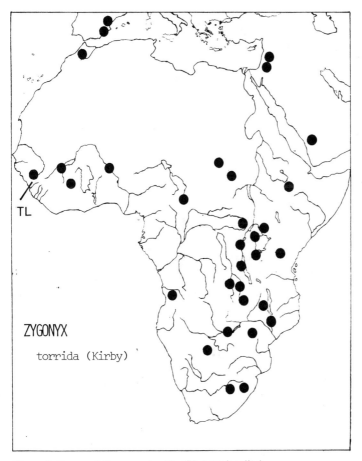

Figure 6. Range of *Zygonyx torrida* (TL type locality).

Some 38 species are found within the confines of the Sahara, or have once crossed the desert leaving no relict populations within it, but along its fringes. All species of this second type have their main range in the Africotropical region, with isolated occurrences in the Maghreb and/or Iberian Peninsula, the Levant, and the major Mediterranean islands of Sardinia, Corsica, and Sicily.

Of the two recent suborders, the Zygoptera are mostly small species and weak flyers, dispersing slowly and over short distances. Adults usually do not

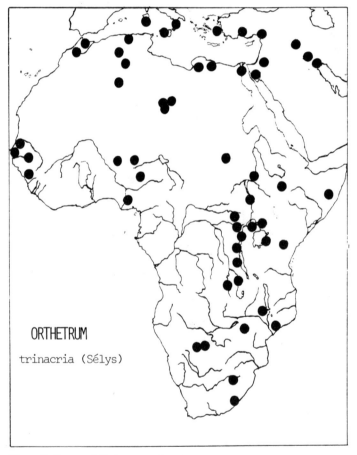

ORTHETRUM

trinacria (Sélys)

Figure 7. Range of *Orthetrum trinacria.*

venture any appreciable distance away from the water in which their larva developed. No cases are known in which the larva estivates, i.e. all Zygoptera are linked to permanent water. Only one species, *Ischnura saharensis* (Ag.) (Figure 1), which has also managed to conquer the whole Sahara, is at times encountered on semi-permanent gueltas (= lakelets). However, in all cases I have on record (Aïr mountains and Tassili-n-ajjer), a permanent guelta was always to be found at a distance of not more than, say, ten kilometers.

Zygoptera are thus good indicators of permanent water. If relict populations are found within the Sahara, they must have been in continuous existence, at least since the latest pluvial spell.

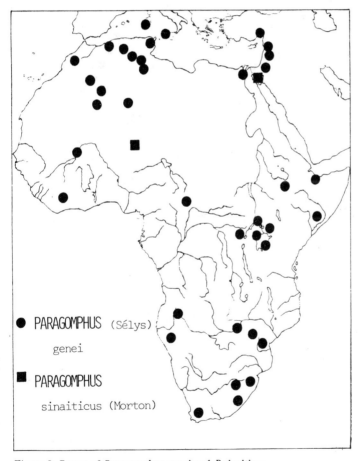

Figure 8. Range of *Paragomphus genei* and *P.sinaiticus.*

9

The suborder Anisoptera is composed of strongly built species, and all are better flyers than the Zygoptera. However, only a minority is migratory (five species: two Libellulidae, three Aeschnidae). Only one species, *Hemianax ephippiger* (Burm.), undertakes true mass migrations over the Sahara, whereby it can remain on the wing during the night. It is thus capable of travelling over intercontinental distances. Although it occurs over most of the African continent, it is preadapted to life in a desert environment by ovipositing in temporary pools and by having a short larval development (about three months) that is tuned to the lifetime of the ponds in which the larvae live. Species like *H.ephippiger* are absolutely useless for tracing back palaeoecological conditions.

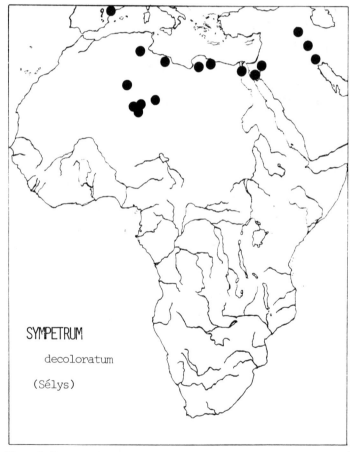

Figure 9. Range of *Sympetrum decoloratum* (?: uncertified record).

10

Another preadaptation to rare and random precipitation was discovered in two members of the Libellulidae: *Orthetrum chrysostigma* (Burm.) and *Trithemis arteriosa* (Burm.). Digging the bed of a temporary guelta in the Aïr in spring 1977 revealed living larvae of *T.arteriosa* in damp sand at a depth of ca. 30 cm. A clump of about 30 larvae of *O.chrysostigma* was found in similar conditions in moist sand at the edge of guelta Bei-Bei (Tassili-n-ajjer). This form of estivation greatly enhances the dispersive potential of the species involved (figure 2): they are ubiquitous over the desert, wherever surface water, permanent or ephemeral, is to be found.

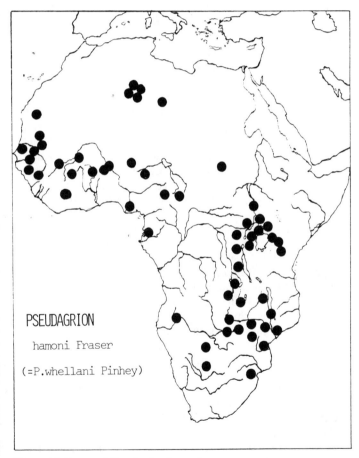

PSEUDAGRION

hamoni Fraser

(=P.whellani Pinhey)

Figure 10. Range of *Pseudagrion hamoni.*

11

Figure 1, beside the distribution of *I.saharensis,* also shows the range of *I.genei* (Ramb.), an endemic of Malta, Sardinia, Corsica, Sicily, and Elba, i.e. occurring on all major central-mediterranean islands and perhaps even in Calabria. The morphology of both species is so very much alike that a common ancestor in the recent past (see further) must be postulated. *I.saharensis* does not occur on the Iberian Peninsula, nor does it cross the Atlas mountains. North-east of this chain it is replaced by the Iberian *I.graellsi.* It is also wanting in the Levant. In the latter area, it is replaced by *I.senegalensis* (Ramb.), which inhabits most of Africa and even extends into Asia. This is a species that has taken advantage of the Nile River as a pathway for dispersal. However, it is not necessarily true that all species that have travelled along the Nile have failed to produce relicts in the Sahara. *Trithemis annulata* (P.de B.) has left a host of relicts in the central Sahara and reaches, via the Oued Saoura, as far north as South Morocco, Tunisia and the Mediterranean islands (figure 3). It has, however, not reached western Morocco and the Iberian Peninsula.

On the other hand, those African species that do occur on the Iberian Peninsula, consistently fail to appear on the Mediterranean islands. Examples are *Diplacodes lefebvrei* (Ramb.) (figure 4), *Brachythemis leucosticta* (Burm.) (figure 5) and *Zygonyx torrida* (Kirby) (figure 6). The reverse (African species with relicts in the Sahara and on the Mediterranean islands but not on the Iberian Peninsula) is again illustrated by *Orthetrum trinacria* (Selys) (figure 7) and by the gomphid *Paragomphus genei* Selys (figure 8). Figure 8 also features *P.sinaiticus* (Morton), a little known species, recorded from Sinai mountains and from Air only. It is possibly widespread in the Arabian mountains, and an example of Bodemheimer's Asiatic invasion fauna. A good case of the latter is offered by *Sympetrum decoloratum* Selys (figure 9), but it should be stressed − as was indicated by *I.senegalensis* − that this movement really worked in both directions.

A significant pattern is further offered by *Pseudagrion hamoni* Fraser (figure 10): its main range covers most of Africa south of the Sahara, and a string of relicts is surviving in Tassili-n-ajjer and in the Fezzan. Like many thermophilous species (more examples will appear among the vertebrates), it has not managed to maintain itself in the high mountains of the Central Sahara and thus, a forteriori, not in the Atlas range. However, its absence from more southerly mountain areas such as the Adrar des Iforhas and the Air cannot be explained in the same manner (see further).

AMPHIBIA

For unknown reasons, no Urodela occur in Africa south of the Sahara. No relict populations from the desert are known either, even though three species occur in the Atlas mountains.

The Anura are represented by ten species. One (*Bufo vittatus* Degens), of Africotropical origin, is restricted to the Nile valley and useless for my present purpose. *Bufo mauretanicus* Schlegel has a much wider distribution: the Maghreb, Rio de Oro (Dumont 1979), Mauretania (Dekeyser & Villiers 1952), but also the valley of the river Niger near Bourem (Guibé 1950), and Agades south of the Aïr (Angel & Lhote 1938). No relicts have been reported from the central or eastern Sahara and the species is not known from the Nile valley. Such a pattern of distribution is of little help in understanding the dynamics of climate change either. *B.mauretanicus* might be a west Mediterranean endemic that possibly originated in the Atlas range, but too little is known about its ecology (are there adaptations to drought?) to speculate further about it.

Since aridity is one of the dominant features of large parts of West, East and Southern Africa, it is no wonder that adaptations to periodical drying-up of aquatic biotopes have evolved, even in the Amphibia. Three Sahara species, all of African origin, have thus learned to obviate the hazards of a drought by digging deep into mud or sand. The animals concerned are *Bufo regularis* (today considered rather a complex of species), *Bufo pentoni* And., and *Tomopterna cryptotis* (Boul.) (= *Pyxicephalus delalandei* Tschudi). As soon as rainfall comes, they emerge from the river beds, often in extremely large numbers, as I observed in some koris (= rivers) of the Aïr in September 1978. Nocturnal habits of the adult, extremely rapid hatching of the eggs following oviposition in ephemeral pools (up to less than one day), extremely rapid larval development (often not longer than some ten days), and great tolerance of the tadpoles to both high water temperatures (up to 30°C) and salinities, are further adaptations that enable the species concerned to thrive under desert conditions. Active dispersal without permanent or continuous water is possible. These species are found in Mauretania (Tagant, Assaba, and Adrar), in the Adrar des Iforhas, and in the Aïr, but not in the Hoggar or in the Tassili-n-ajjer. The only record from central Sahara known to me is that of *Bufo regularis* in the lowland oasis of Ghat (Libya) (Scortecci 1937). Even more than the dragonfly Pseudagrion, these Anura are unable to stand the low winter temperatures of the central Saharan mountains, despite the availability of much surface water. Ghat (and in fact quite a few other lowland oases of the Fezzan) make an exception because they offer a much milder microclimate.

One Africotropical Anuran that has left a relict population in the Tassili-n-ajjer (a lower and less selective environment than the Hoggar) is *Bufo mascareniensis* (D. & B.) (*guelta Idefil:* Pellegrin 1921). This species is much more tightly linked to water than the preceding, and it is therefore a valuable relict. *Ptychadena tigrina occipitalis* (Günther), of similar ecology and range, is found in the South-west (as far north as the Adrar of Mauretania), and in the oasis of Ghat (Fezzan) (cfr. *Bufo regularis*). It makes another valuable relict. The two dominant Anura of both the Hoggar (and annexes: Ahnet, Mouydir and Tefedest) and Tassili-n-ajjer are of palaearctic origin. *Rana ridibunda perezi* Seoane and *Bufo viridis* Laurenti are widespread in the Maghreb countries and are linked to the central Saharan populations by a string of relicts in the valley of the Oued Saoura and in the oases of the plateau of Tidikelt. Both are restricted to permanent water. They are, however, not found

13

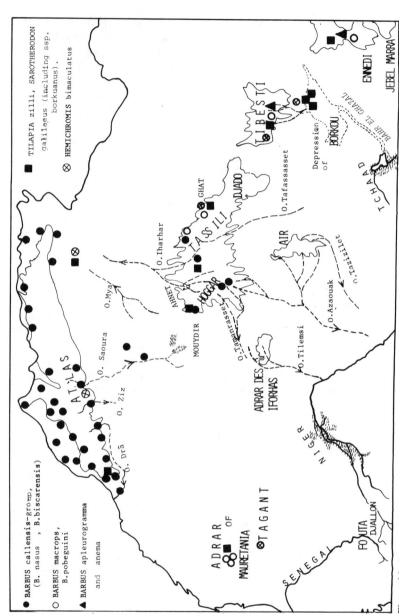

Figure 11. Ranges of some fish species in the Sahara and the Maghreb.

14

in the presently permanent gueltas of Aïr and Adrar des Iforhas, nor in Tibesti and Ennedi. The Ennedi plateau, which is even more strongly influenced by the summer monsoon than the southern Aïr, is the only site where the tropical *Xenopus mülleri* (Peters) has been found (Monod 1968).

REPTILES

The only case – albeit a spectacular one – of an aquatic reptile with Saharan relicts is the Nile crocodile, *Crocodylus niloticus* Laurenti. A population survived in the large guelta of Iherir (Tassili-n-ajjer) until 1920. However, more populations had existed in Tunisia and the Cirenaica until Roman times, and skeletons in the bed of the Oum Aghouaba lakes, north of Atar (Adrar of Mauretania), have been C14 dated at not more than 2,000 BP (Chamard *et al.* 1970). Today, relict populations persist in the Tagant (Mauretania), in Tibesti, and in the Ennedi. No protohistorical skeletons have been recorded from Adrar des Iforhas or from Aïr.

FISH

Eighteen species and a possible subspecies have been recorded from the Sahara (table 1) (figure 11). Among these, seven are restricted to the Tibesti or to the Tibesti-Ennedi-Borkou area (Daget 1959). Although some endemism has been claimed for the ichtyofauna of Tibesti, it was shown (Thys van den Audenaerde 1970) that *Tilapia borkuana* is, at best, a melanic race of *Tilapia* (= *Sarotherodon*) *galilaea* Artedi. The second so-called endemic, *Labeo tibestii* Pell. might in turn be a synonym of *L.annectens* Boul. from the River Niger. This reduces all Tibesti-Ennedi relicts to an extract of the Nile-Chad-Niger fauna. The very close relationship between the ichtyofauna of the Nile and the Niger is well known. Greenwood (1976) states that of the 115 species of fishes found in the Nile, at least 74 are shared with the 120-130 species of the Niger, and ten nilotic species do, on top of this, not cross the intermediate site of Lake Chad. Recent connections between all three basins must thus have existed. The pathway from Nile to Chad is the Bahr el Ghazal lying between the Tibesti and Ennedi (figure 11). It functioned at very high levels of Lake Chad, when the lake was in fact reaching the foot of both mountain areas and fish could invade them. Data by Servant (1973), show that the most recent high level of Lake Chad was around 5,400 BP. This means that maximum isolation time of the Tibesti populations is of that order or magnitude. As will appear further, 5,000 years are insufficient to allow endemisms to develop with a good probability of success.

The Atlas mountains house an endemic group of Barbus. Although their ultimate taxonomical status is still a matter of dispute Pellegrin (1921, 1939)

Table 1. The distribution of fishspecies in the Sahara.

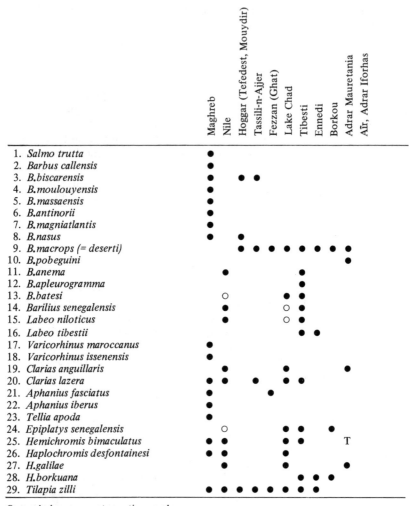

	Maghreb	Nile	Hoggar (Tefedest, Mouydir)	Tassili-n-Ajjer	Fezzan (Ghat)	Lake Chad	Tibesti	Ennedi	Borkou	Adrar Mauretania	Aïr, Adrar Iforhas
1. *Salmo trutta*	●										
2. *Barbus callensis*	●										
3. *B.biscarensis*	●		●	●							
4. *B.moulouyensis*	●										
5. *B.massaensis*	●										
6. *B.antinorii*	●										
7. *B.magniatlantis*	●										
8. *B.nasus*	●		●								
9. *B.macrops (= deserti)*			●	●	●	●	●	●	●	●	●
10. *B.pobeguini*											●
11. *B.anema*			●				●				
12. *B.apleurogramma*							●				
13. *B.batesi*			○				●	●			
14. *Barilius senegalensis*			●			○	●				
15. *Labeo niloticus*			●			○	●				
16. *Labeo tibestii*							●	●			
17. *Varicorhinus maroccanus*	●										
18. *Varicorhinus issenensis*	●										
19. *Clarias anguillaris*			●				●				●
20. *Clarias lazera*	●	●			●		●	●			
21. *Aphanius fasciatus*	●					●					
22. *Aphanius iberus*	●										
23. *Tellia apoda*	●										
24. *Epiplatys senegalensis*			○				●	●		●	
25. *Hemichromis bimaculatus*	●	●					●	●			T
26. *Haplochromis desfontainesi*	●	●					●				
27. *H.galilae*		●					●				●
28. *H.borkuana*								●	●	●	
29. *Tilapia zilli*	●	●	●	●	●	●	●	●			

Open circles represent question marks
T : Tagent

distinguishes as many as 13 species and four 'varieties', while Estève (1947) reduces all to the couple *Barbus callensis* Cuv. and *B.nasus* Günther, and Almaça (1970) distinguishes the seven species listed in table 1, it is clear that the Maghreb group is fairly distant from the circa 250 Africotropical Barbus-species, and the same holds true for the two Varicorhinus species of Morocco. Relict populations of both *B.biscarensis* Cuv. and *B.nasus* Günther are known from the Saoura valley, the Hoggar and annexes (e.g. Tahount Arak), and the Tassili-n-ajjer. Conversely, two equatorial Barbus-species, *B. macrops* Boul. (= *B.deserti* Pell.) and *B.pobeguini* Pell., are found in numerous populations, not only in the Adrar of Mauretania, Ennedi, and Tibesti, but also in the Tassili-n-ajjer. The same applies to the cichlid species that have Saharan relicts (*Sarotherodon galilaeus* and *Tilapia zilli*, and *Hemichromis bimaculatus* Gill.), with the additional remark that all except *S.galilaeus* also reach the southern slopes of the Atlas mountains (figure 11). A fourth ciclid (*Haplochromis desfontainesi*) is only found in the latter area as a relict, and not in Sahara itself.

Two African species of catfish are found in the Sahara (figure 12): *Clarias anguillaris* L. occurs in the West, reaching the Adrar of Mauretania; *C.lazera* C. & V. lives in the Ennedi, Tibesti, Tassili-n-Ajjer, Fezzan (gueltas in the bed of Oued Tichamallt, O.Tarat, O.Iseien), and reaches the foot of the Atlas in the Oasis of Tolga near Biskra (Algeria). *C.lazera* has also reached the Levant, and is of limited economical importance in Lake Kinneret.

Catfishes are preadapted to life in arid environments by their ability for air-breathing, permitting them to migrate (and even to feed) on land in a damp atmosphere. They can survive not too long periods of drought in all but the dryest mud.

Clarias gariepinus, a species from Southern Africa, that has not been found in Sahara, but is very closely related to *C.lazera,* is known to occur in inter-mittent streams in Zimbabwe. It survives periods of drought at a depth of as much as 3 m in the river bed. In the same area, an identical phenomenon has been recorded in the cichlid *Sarotherodon mossambicus* (Donnelly 1978). We do not know exactly how long these fishes can survive under these circum-stances, although a month or so might be a reasonable estimate.

Further, there is no direct information on possible similar performances in the Saharan cichlids, but the possibility should be taken seriously. Circumstan-tial evidence from personal observations is the long survival time outside water of net-captured *S.galilaeus* (several hours), while at guelta Efenni (Tassili-n-ajjer), I noted that *T.zilli* is capable of moving between a series of shallow, iso-lated rockpools by jumping out of the water and across bare rock over dis-tances of several metres.

No such adaptations are known in Barbus. Quite to the contrary, I observed mass-mortality in *B.pobeguini* in the shallow guelta of Hamdoun, Adrar of Mauretania, as water temperatures rose to 30°C.

None of the adaptations cited will, however, permit fish to survive year-long periods of severe drought. The complete absence of fish and the ecologi-cal nature of the Amphibia (sand-burrowing species) in both the Air and the Adrar-des-Iforhas has therefore two implications (Dumont 1978): a hyperarid spell in the recent past lasted long enough to eliminate all species from that area, and it was not followed by a period humid enough to permit recolonisa-tion.

FURTHER INTERPRETATIONS OF THE PATTERNS OF DISTRIBUTION IN TERMS OF CLIMATIC FLUCTUATIONS SINCE THE LATE WÜRM

The differential and mutually exclusive pattern seen in many Odonata appears to me to be directly related to the last stadial of the Würm, corresponding to a maximum extension of the ice-sheet over Europe about 20,000 BP. A 130 m eustatic lowering of the sea level narrowed the distance between Tunisia, Corsardinia and Italo-Sicily to such an extent that quite a few species had little difficulties in crossing this stepping stone. However, some amount of filtering

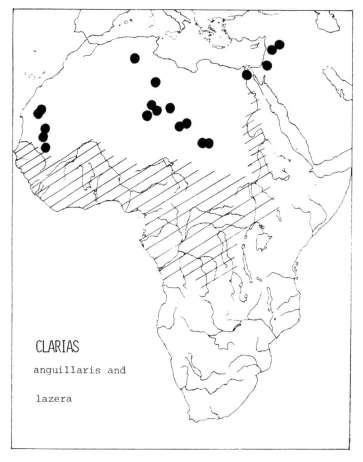

CLARIAS

anguillaris and

lazera

Figure 12. Range of both *Clarias anguillaris* and *C.lazera*. Dashed area: main range.

Figure 13. Estimated faunal movements shortly before 20 000 BP; at the shores of the Mediterranean, around 18 000 BP. Dashed or stippled areas are glaciated.

is evident in this faunal exchange: European species crossed from north to south, but the Ethiopian dragonflies that migrated north were all species tolerant of wide fluctuations in temperature, and no African fish or amphibian succeeded in getting established on Sardinia or Sicily. The reason for this is probably the cold climate that reigned over the Mediterranean at that time (most of Italy was covered by a steppe), and selected strongly against thermophilous species. In addition, there was aridity, not only in the Sahara but over much of Africa (Burke *et al.* 1971), making the Ethiopian migrants themselves relics of an earlier moist phase, to be situated around 30,000-25,000 BP.

That African species could not colonize the Iberian Peninsula was a consequence of the much more extensive glaciation on the Atlas mountains, blocking the passage across Gibraltar hermetically. Some faunal exchange with Arabia and the Levant may have been possible (figure 13).

After 18,000 BP, with deglaciation proceeding, and the climate warming up, the Atlas barrier gradually waned. At the same time, the sea level started rising, to exceed present levels by about 3 m around 5,000 BP; corresponding to the Nouakchott transgression (Elouard 1968; see also Delibrias 1973) and the so-called climatic optimum in Europe. In the course of these events, crossing the central Mediterranean became impossible but the route to the Iberian Peninsula opened, and a new migration wave started (figure 14). In all probability, it coincided with the warm-humid spell over both the great East-African lakes and the Sahara (ca. 12,000-7,000 BP). Since the Sinai and the Negev deserts also became relatively humid, considerable exchange between the Sahara, the Nile valley, the Levant, and, near the horn of Africa, between North-east Africa and Southern Arabia occurred. Unfortunately, too little is known about the aquatic fauna of Arabia to discuss the latter possibility in any detail. Doubtlessly, this period was of the greatest importance in shaping the patterns of animal distribution that today we witness in the form of relicts, but it also appears that a period of isolation lasting some 7,000 years is not sufficient to allow speciation to proceed. Not a single endemic species has indeed evolved since.

Even the older migration wave has produced little speciation. Only among dragonflies is there one notable exception: *Ischnura saharensis* has evolved allopatrically from *I.genei*, but the relationship between both is still extremely close. It seems that the random process of speciation might require a minimum period of ca. 20,000 years.

A general drought around 7,000-6,000 BP brought about the relictisation of most of the immigrants of the Sahara. It was followed by the well-documented neolithic humid spell (ca. 6,000-3,000 BP). In an earlier contribution from which I also borrowed earlier in this paper, I argued that this was restricted to the northern and central Sahara. During this whole period, or a significant portion of it, the Southern Sahara and Central Sahel were experiencing a climate, sufficiently arid to eliminate all faunas linked to permanent water.

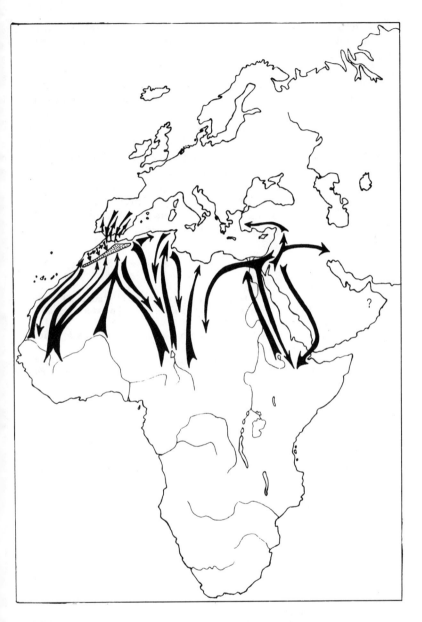

Figure 14. Faunal movements around 12 000-8 000 BP. One branch of the oriental invasion fauna, possibly reaching the Maghreb and Spain, not drawn on the figure.

21

A later mitigation of this climate, corresponding to the onset of modern conditions, did not create a consistent water-link between, say, the Aïr and the Niger, so that recolonisation could never occur.

The aquatic dragonflies that did recolonize the Aïr gueltas, are palearctic or oriental in nature, or are wide-ranging migrants. The absence of neolithic artwork from the Aïr proper (not the adjacent Tenere) was used as an argument for dating this period, but there is additional evidence in studies on remobilisation of old dunes in the Chad area (Servant 1973), in northwestern Nigeria (Sombroek & Zonneveld 1971), and in Western Sudan (Wickens 1975). In all these cases, strong eolian activity was noted around 5,000-4,000 BP.

It is also clear why Tibesti and Ennedi have not lost their fish and amphibia like the Aïr: the transgressions of Lake Chad, depending on precipitation deep in the equatorial zone of West-Africa, were independent of the climate near the lake. The same applies to the fauna of the Adrar of Mauretania, which may have been replenished by monsoonal transgressions of the River Senegal.

REFERENCES

Almaça, C. 1968 (1969). Révision critique de quelques types de Cyprinidés d'Europe et d'Afrique du nord des collections du muséum national d'histoire naturelle. *Bull. Mus. nat. Hist. nat., Paris* S2, 40: 1116-1144.

Almaça, C. 1970. Sur les barbeaux (genre et sousgenre Barbus) de l'Afrique du Nord. *Bull. Mus. nat. Hist. nat., Paris* S2, 42: 14-158.

Almaça, C. 1971. Sur la collection de Barbeaux (genre et sous-genre Barbus) ibériques et nord-africaines du Forschungsinstitut Senckenberg (Frankfurt AM). *Arq. Mus. Bocage* S2, 3, 21: 5pp.

Angel, F. & H.Lhote 1938. Reptiles et Batraciens du Sahara central et du Soudan. *Bull. Com. Et. Hist. Sci. AOF* 21: 345-384.

Bodenheimer, F.S. 1938. On the presence of an Irano-Turanian relict fauna in North-Africa. *Mém. Soc. Biogéogr.* 6: 67-79.

Burke, K., A.B.Durotype & A.J.Whiteman 1971. A dry phase south of the Sahara 20 000 years ago. *W.afr. J. Archaeol.* 1: 1-8.

Butzer, K.W. 1971. Quartäre Vorzeitklimate der Sahara. In: H.Schiffers (ed.), *Die Sahara und ihre Randgebiete*, I. Weltforum, München: 349-388.

Chamard, P., R.Guitot & G.Thilmans 1970. Le lac holocène et le gisement néolithique de l'Oum Arouaba (Adrar de Mauritanie). *Bull. IFAN* B32: 668-740.

Daget, J. 1959. Note sur les poissons du Borkou-Ennedi-Tibesti. *Trav. Inst. Rech. Sah.* 18: 173-181.

Dekeyser, P.L. & A.Villiers 1956. Contribution à l'étude du peuplement de la Mauritanie. Notations écologiques et biogéographiques sur la faune de l'Adrar. *Mém. IFAN* 44: 222pp.

Donnelly, B.G. 1978. Evidence of fish survival during habitat dessication in Rhodesia. *J. Limnol. Soc. sthn.Afr.,* 4: 75-76.

Dumont, H.J. 1978. Neolithic hyperarid period preceded the present climate of the Central Sahel. *Nature* 274: 356-358.

Dumont, H.J. 1979. Limnologie van Sahara en Sahel. Unpublished thesis, University of Gent. 557pp.

Dumont, H.J. 1981. Zooplankton and the Science of biogeography: The example of Africa. In: W.C.Kerfoot (ed.), *Evolution and Ecology of Zooplankton Communities.* University Press of New England.

Elouard, P. 1968. Le Nouakchottien, étage du quaternaire de Mauritanie. *Ann. Fac. Sci. Dakar,* 22: 121-138.

Estève, M. 1947. Etude biométrique des barbeaux maroccains. *Bull. Mus. nat. Hist. nat., Paris* S2, 19: 265-270.

Estève, M. 1949. Poissons du Sahara Central. *Bull. Soc. zool. France* 74: 19-20.

Estève, M. 1952. Poissons de Mauretanie et du Sahara orientale. Un nouveau sous-genre de Barbus. *Bull. Mus. nat. Hist. nat., Paris* S2, 24: 176-179.

Faure, H. 1969. Lacs quaternaires du Sahara. *Mitt. int. Ver. Limnol.* 17: 131-146.

Gauthier, H. 1938. La vie désertique dans les déserts subtropicaux. In: La vie dans la région désertique nord-tropicale de l'Ancien-Monde. *Mém. Soc. Biogéogr., Paris* 6: 107-120.

Greenwood, P.H. 1976. Fish fauna of the Nile. In: The Nile biology of an ancient river (J.Rzoska, ed.). *Monogr. Biol.* 29: 127-141.

Guibé, J. 1950. Batraciens. In: Contributions à l'étude de l'Aïr. *Mém. IFAN* 10: 329-330.

Lavauden, L. 1926. Les Vertébrés du Sahara. *Guénard, Tunis* 200pp.

Monod, T. 1938. Remarques générales. *Mem. Soc. Biogéogr.* 6: 375-405.

Monod, T. 1968. Contribution à l'étude des eaux douces de l'Ennedi. III. Crustacés décapodes. *Bull. IFAN* A30: 1350-1353.

Niethammer, G. 1971. Die Fauna der Sahara. In: H.Schiffers (ed.), *Die Sahara und ihre Randgebiete,* vol.1: 499-587. Weltforum, München.

Pellegrin, J. 1909. Description d'un Barbus nouveau du Sahara. *Bull. Mus. hist. nat. Paris* 153: 972-974.

Pellegrin, J. 1911. Les Vertébrés aquatiques du Sahara. *CR Acad. Sci., Paris* 153: 972-974.

Pellegrin, J. 1913. Les Vertébrés des eaux douces du Sahara. *Cr. Ass. Fr. Avanc. Sci., Congrès de Tunis* 1913: 346-352.

Pellegrin, J. 1921. Les poissons des eaux douces de l'Afrique du Nord Française. *Mém. Soc. Sci. nat. Maroc.* 1(2): 1-216.

Pellegrin, J. 1927. La présence du Crapaud vert dans le Hoggar. *Bull. Soc. centr. aqinc.* 34: 120-121.

Pellegrin, J. 1929. Mission saharienne Augiéras-Draper 1927-28. Poissons. *Bull. Mus. nat. Hist. nat.,* 2, 1: 134-139.

Pellegrin, J. 1931. Reptiles, Batriciens et Poissons du Sahara central recueillis par le Prof. Seurat. *Bull. Mus. Hist. nat.* 2, 3: 216-218.

Pellegrin, J. 1933. Poissons. Voyage de Ch.Alluaud et P.A.Chappuis en Afrique Occidentale française. *Arch. hydrobiol.* 26: 101.

Pellegrin, J. 1939. Les Barbeaux de l'Afrique du nord française: description d'une espèce nouvelle. *Bull. Soc. Sci. nat. Maroc.* 19: 1-10.

Quézel, P. 1965. La végétation du Sahara. Du Tchad à la Mauritanie. *Geobotanica selecta* 2: 333pp., Fischer, Stuttgart.

Quézel, P. 1971. Flora und vegetation der Sahara. In: H.Schiffers (ed.), *Die Sahara und ihre Randgebiete,* vol.1: 429-475. Weltforum, München.

Schnurrenberger, H. 1962. Fishes, amphibians, and reptiles of two Libyan Oases. *Herpetologica* 18: 270-273.

Scortessi, G. 1937. La Fauna. In: Il Sahara Italiano. Parte prima. Fezzan e Oasi di Gat. *Reale Acad. Ital., Roma* 211-239.

Servant, M. 1973. *Séquences continentales et variations climatiques: évolution du bassin du Tchad au Cénozoique supérieur.* DSc Thesis, Paris: 348pp.

Sombroek, W.G. & I.S.Zonneveld 1971. Ancient dune fields and fluviatile deposits in the Rima-Sakoto basin (N-W Nigeria). *Netherlands Soil Survey Inst., Wageningen* Paper no.5, 109pp.

Talbot, M.R. 1980. Environmental responses to climatic changes in the West African Sahel over the past 20 000 years. In: M.J.A.Williams & H.Faure (eds.), *The Sahara and the Nile* pp.37-62. Balkema, Rotterdam.

Thys van den Audenaerde, D. 1970. *Bijdrage tot een systematische en bibliografische monografie van het genus Tilapia (Pisces, Cichlidae).* Doktoraatsthesis, Gent, 261pp.

Van Zinderen Bakker, E.M. & J.A.Coetzee (eds.) 1980. *Palaeoecology of Africa 12.* A.A. Balkema, Rotterdam. 408pp.
Wickens, G.E. 1975. Changes in the climate and vegetation of the Sudan since 20 000 BP. *Cr. VIII° Réun. AETFAT, Génève 1974. Bossiera* 24a: 43-65.
Williams, M.A.J. & H.Faure (eds.) 1980. *The Sahara and the Nile. Quaternary environ- ments and prehistoric occupation in northern Africa.* A.A.Balkema, Rotterdam. 607pp.
Zavattari, E. 1934. La Fauna ittica di Gat (Fezzan) e la affinita zoogeografica del terri- torio di Gat con il Sahara algerino. *Rendic. r. Ist. lomb. Sci. Lett.* 2, 65: 75-82.

CHRONOLOGY OF GLACIAL AND PERIGLACIAL DEPOSITS, MOUNT KENYA, EAST AFRICA: DESCRIPTION OF TYPE SECTIONS

W. C. MAHANEY

Geography Department, Atkinson College, York University, Downsview, Ontario, Canada

ABSTRACT

During the Quaternary valley glaciers on Mount Kenya, radiating from a central ice cap, descended to 2 900 m. A five-fold sequence of tills, and associated periglacial deposits, are present in most drainages; all glacial deposits have been named from type localities on the western and northern flanks of the mountain. Field relationships indicate that the Höhnel Diamicton is of glacial origin and of probable early to mid-Quaternary age. Teleki till, deposited prior to the last glacial maximum (Würm, Wisconsinan) is found in lower glaciated valleys at 2 900 m. These deposits carry well-developed soils, and at the type locality (site TV23) a surface soil overlies a buried paleosol formed in older drift that may correlate with the Höhnel Diamicton. Younger glacial deposits, presumed coeval with the last glacial maximum, are here termed Liki till. These deposits mantle most valleys above 3 200 m and are divisible into early and late stages on the basis of topographic position. In the high valleys deposits of Liki-II till are found above 3 900 m, and date at 12 590 ± 300 BP. In the lower valleys Liki-I drift has not been radiocarbon dated. Younger till of Neoglacial age is considered late-Holocene in age (< 1 000 BP) on the basis of relative soil development and weathering features.

INTRODUCTION

Mount Kenya is distinctive in that some of the older tills are related to volcanic rocks and organic materials that can be dated by radiometric methods. In addition, relative age-dating methods have been used to demonstrate age differences in the till sequence (Mahaney 1972, 1979). Such relative dating methods include various rock-weathering parameters, loess thickness, vegetation cover (including lichen data), and soil properties. Recent reviews of these methods have been summarized by Mahaney (1979: 166-167).

Valley-to-valley correlation of tills is not easy to achieve as a result of dense forest cover below 3 200 m, and variations in climate between the west and

Table 1. Summary of characteristics useful in subdividing and correlating deposits, Mt Kenya, East Africa

Deposit	Elevation (m)	Vegetation	Boulders 10 m²	Weathering ratios*			Weathering rinds**		
				% fresh	% Wx	n	Average max. (mm)	Average min. (mm)	n
Lewis Till	4 550	Upper Afroalpine	120	100	0	100	1.0	nil	100
Tydall Till	4 350	Upper Afroalpine	102	100	0	100	3.7	0.02	100
Liki-II Till	3 990	Upper Afroalpine	42	98	2	100	4.6	1.3	100
Liki-I Till	3 260	Ericaceous Zone	30	95	5	100	7.0	1.5	100
Teleki Till	2 990	Bamboo Forest	6	52	48	100	15.6	2.9	100
Höhnel Diamicton	4 100	Upper Afroalpine	–	–	–	–	–	–	–

* Weathered-fresh differentiation based on surface state of the boulder as generally unweathered (e.g. fresh) or weathered (Wx), cavernous, and rotten.

** Oxidation rinds are developed to non-uniform depths in clasts found on moraine surfaces. The maximum rind is the maximum depth of discoloration measured on clasts split with a hammer. The minimum rind is the minimum depth.

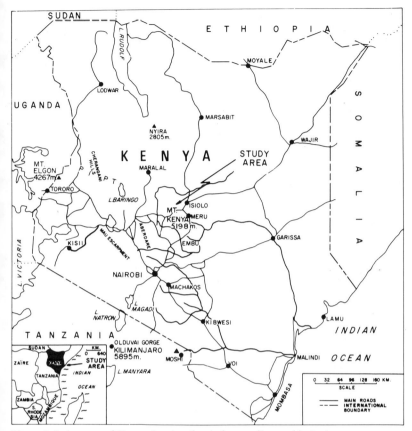

Figure 1. Index map of Kenya showing the study area discussed in text.

north flanks of the mountain. This paper outlines the general distribution and dimensions of Quaternary glaciers in the Mount Kenya area and discusses the type section for each stratigraphic unit. Problems associated with the identification and differentiation of each unit and their age relationships are also discussed.

DIAMICTONS IN INTERFLUVES

In a few localities, deposits considered to be tills, are located in interfluves 25-125 m above existing drainages. The major deposits are above Lake Höhnel on the Höhnel/Teleki interfluve and along the west fork of the Kazita River (Baker 1967; Mahaney 1979). Correlation of these diamictons will have to

27

be based on similar topographic positions and K/Ar dates. The diamicton in the Höhnel/Teleki interfluve (figure 2) is located 125 m above the Teleki Valley. It contains striated and polished material of boulder, cobble, and pebble sizes, indicating a probable glacial origin. This deposit varies from 2 to 0.5 m in thickness, is discontinuous, thinning out laterally, towards the east. Underlying and overlying volcanic rocks are presently being dated to provide bracketing maximum and minimum ages. Additional field work is required to determine if other diamicts are present on Mount Kenya or Ithanguni, a parasitic cone to the northeast.

TILLS WITH POORLY PRESERVED MORAINE FORM

Unlike the older diamictons, tills deposited during the Teleki Glaciation are closely related to the valley floor over which the ice advanced. At the type locality for Teleki till (site TV23; location figure 2) surface drift overlies older drift that may be coeval with the Höhnel Diamicton. Surface weathering data for Teleki till (table 1) indicate an advanced state of weathering with a low boulder frequency, relatively high percentage of weathered stones, and fairly large rinds. Multi-storey paleosols in site TV23 are described as follows:

TV23: post-Teleki relict paleosol and pre-Teleki buried paleosol (location figure 2)

Horizon	Depth (cm)	Description*
AII	0-18	Brownish black (10YR 2/3m; 3/3d) and dark brown (10YR 3/3d) silty clay loam, granular structure, friable, slightly plastic, and slightly sticky.
IIA12	18-38	Brownish black (10YR 3/2m) and dull yellowish brown (10YR 4/3d) clay, granular structure, firm, plastic and sticky.
IIB21	38-91	Dark reddish brown (5YR 3/3m) and brown (7.5YR 4/4d) clay, blocky structure, firm, plastic, and sticky.
IIB22	91-104	Brown (7.5YR 4/4m) and dull yellowish brown (7.5YR 5/4d) clay, blocky structure, firm, plastic and sticky.
IIC1ox	104-168	Reddish brown (5YR 4/6m), bright reddish brown (5YR 4/5m) and orange (7.5YR 6/6d) clay, massive structure, friable, plastic, and slightly sticky.
IIC2ox	168-176	Dull reddish brown (5YR 4/4m) and bright brown (7.5YR 5/6d) clay, massive structure, firm, plastic, and sticky.
Ab	176-186	Black (10YR 1.7/1m), yellowish brown (10YR 5/6m) and dull brown (7.5YR 5/3d) loam, massive structure, firm consistence, plastic, and sticky.
B21b	186-199	Yellowish brown (10YR 5/6m), black (10YR 1.7/1m) and dull yellow orange (10YR 6/4d) clay, massive structure, firm, plastic, and sticky.
IIB22b	199-217	Brown (10YR 4/4m) and dull yellow orange (10YR 7/3d) clay, massive structure, firm, plastic and very sticky.
IIB23b	217-247	Brown (7.5YR 4/4m; 10YR 6/4d) and dull yellow orange (10YR 6/4d) clay, massive structure, firm, plastic, and sticky.

Horizon	Depth (cm)	Description*
IIC1oxb	247-277	Brown (10YR 4/6m) and bright yellowish brown (10YR 6/6d) clay, massive structure, firm, very plastic, and sticky.
IIC2oxb	277+	Brown (10YR 4/6m), some light gray (10YR 8/2m) and dull yellow orange (10YR 6/4d) clay, massive structure, very firm consistence, very plastic and sticky.

* Colors are from Oyama and Takehara (1970) and are given as moist (m) and dry (d).

Sola is buried and relict paleosols are distinctly different in particle size content (table 2). The buried soil is considerably heavier in texture, with greater amounts of clay and less sand when compared with the relict soil in Teleki till. Particle size distributions (figure 3) show the clear differences between

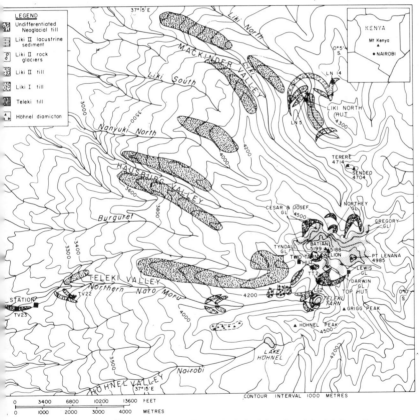

Figure 2. Mount Kenya, with location of principal glacial and periglacial deposits, and type sections described in text.

29

Table 2. Selected physical[1] and mineral[2] properties of Quaternary Type Sections, Mount Kenya, East Africa

Site[3]	Soil horizon	Depth (cm)	Elevation (m)	Vegetation	Age	<2 mm % sand (2 mm - 63 μ)	% silt (63 - 4 μ)	% clay (<4 μ)	Mineralogy K	H	N	I	Q	F	M	G
LG2	C	0-10	4 450	Upper Afro-alpine	post-Neo-glacial soils	75.0	19.2	5.8	—	—	—	—	xx	x	—	—
TT2	A1	0-20	4 350	Upper Afro-alpine	post-Neo-glacial soils	65.5	18.0	16.5	—	—	—	—	tr	tr	—	—
	Cox	20-41				64.1	24.4	11.5	—	—	—	—	tr	tr	tr	—
	Cn	41+				75.3	20.9	3.8	—	—	—	x	X	x	tr	tr
LN14	O2	8-0	3 990	Upper Afro-alpine	Post-Liki-II soil	8.6	48.4	43.0	—	—	—	x	X	x	x	—
	A1	0-10				30.5	28.5	41.0	—	—	—	—	X	tr	—	—
	B2	10-20				33.8	30.2	36.0	—	—	—	—	tr	tr	—	—
	IIC1ox	20-33				67.2	22.3	10.5	—	—	tr	x	X	x	x	—
	IIC2ox	33-58				82.3	8.2	9.5	—	—	—	—	X	x	x	—
	IICn	58+				89.5	3.5	7.0	—	—	—	—	X	x	x	—
TV22	A11	0-15	3 260	Ericaceous Zone	Post Liki-I soil	7.4	62.6	30.0	—	—	—	—	xx	tr	—	—
	A12	15-28				11.6	64.4	24.0	—	—	—	—	xx	x	tr	—
	IIA13	28-43				19.7	45.8	34.5	—	—	—	x	X	tr	—	—
	IIB2	43-56				41.2	42.8	16.0	tr	—	—	tr	tr	tr	?	—
	IICox	56-82				56.5	27.5	16.0	tr	tr	—	—	tr	tr	—	—
	IICn	82+				38.9	45.1	16.0	—	—	—	—	tr	x	—	x
TV23	A11	0-18	2 990	Bamboo forest	Post-Teleki soil	12.3	62.7	25.0	—	—	—	—	X	x	x	—
	IIA12	18-38				11.9	21.6	66.5	—	—	—	—	tr	tr	—	—
	IIB21	38-91				15.5	16.5	68.0	—	—	tr	tr	tr	—	—	—
	IIB22	91-104				11.1	7.9	81.0	—	—	—	tr	tr	tr	—	—
	IIC1ox	104-168				24.6	14.4	61.0	tr	tr	—	tr	tr	tr	—	—
	IIC2ox	168-176				17.5	28.0	54.5	tr	—	tr	tr	tr	—	—	—
	Ab	176-186				45.5	33.5	21.0	tr	tr	x	tr	x	tr	—	—
	B21b	186-199		Pre-Teleki soil		5.9	28.1	66.0	tr	tr	x	tr	x	tr	tr	—
	IIB22b	199-217				1.8	11.7	86.5	—	x	tr	tr	tr	—	.	—
	IIB23b	217-247				2.0	11.0	87.0	—	—	x	tr	tr	—	—	—
	IIC1oxb	247-277				1.1	3.7	95.2	—	x	x	—	—	—	—	—
	IIC2oxb	277+				2.8	8.2	89.0	tr	tr	x	tr	x	tr	—	—

1. Data are given in weight-percentages of sand, silt and clay (<2 mm). Coarse particle sizes (2 000-63 μ) determined by sieving; fine particle sizes (63-1.95 μ) determined by hydrometer.
2. Mineral abundance is based on peak height: nil (—); minor amount (tr); small amount (x); moderate amount (xx); abundant (X). Minerals are kaolinite (K); halloysite (H); nacrite (N); illite (I); quartz (Q); feldspar (F); mica (M); gibbsite (G).
3. Section locations are on figure 2.

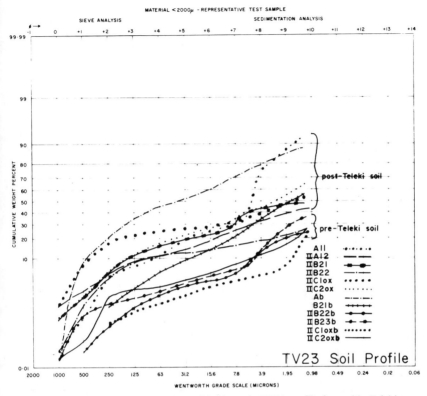

Figure 3. Particle size distributions for soil horizons in TV23 profile formed in Teleki and pre-Teleki tills.

the two soils where the IIB22b, IIB23b, IIC1oxb and IIC2oxb horizons have heavier textures. Calculations of the frequency distribution center of gravity* for the sola in each profile yield values of 8.9 ($\bar{x}$ $\emptyset$) in the relict paleosol, and 12.0 ($\bar{x}$ $\emptyset$) in the buried paleosol. These data show a frequency distribution center of gravity shift of considerable magnitude suggesting considerable differences in time. High silt content in the A11 and Ab/B21b horizons may be of aeolian origin.

Differences between the two paleosols are further supported by mineralogical data (table 2). Within the clay mineral suites, kaolinite, halloysite, nacrite, and illite dominate in trace to small amounts. The presence of nacrite, a

---

* The frequency distribution center of gravity (mean phi) is calculated as follows:

$$\bar{x} \, \emptyset = \frac{25\text{th} + 50\text{th} + 75\text{th percentiles}}{3} \; ; \text{ from figure 3.}$$

Table 3. Selected chemical properties of the < 2 mm fractions of the soil horizons in table 1.

Site*	Horizon	Depth (cm)	pH (1:1)	Extractable cations (meq/100 g) Na⁺	K⁺	Ca⁺⁺	Mg⁺⁺	CEC meq/100 g	Organic matter %	Organic carbon %	Nitrogen %	Fe₂O₃ %	Carbon/nitrogen
LG2	C	0-20	6.8	2.68	0.39	1.20	0.35	5.0	0.7	0.4	nil	0.37	nil
TT2	A1	20-41	5.2	0.22	0.11	0.30	0.08	24.3	3.5	2.0	0.393	0.37	15.3
	Cox	41+	5.8	0.22	0.05	0.35	0.05	19.3	3.4	1.9	0.188	0.47	10.5
	Cn	41+	6.4	0.74	0.25	0.74	0.10	9.6	0.4	0.3	0.011	0.15	23.6
LN14	O2	8-0	5.1	0.23	1.24	0.76	0.90	65.3	38.9	22.6	1.620	0.84	14.0
	A1	0-10	5.0	0.16	0.20	0.10	0.17	57.1	24.4	14.2	0.718	1.10	19.8
	B2	10-20	5.2	0.15	0.19	0.09	0.05	37.7	5.8	3.4	0.279	1.15	12.2
	IIC1ox	20-33	5.5	0.16	0.45	0.80	0.08	18.7	0.7	0.4	0.019	0.34	22.1
	IIC2ox	33-58	5.6	0.20	0.70	1.30	0.12	17.0	0.6	0.3	0.009	0.28	35.6
	IICn	58+	5.6	0.32	1.25	2.40	0.18	18.7	0.5	0.3	0.003	0.22	86.7
TV22	A11	0-15	5.0	1.06	3.32	1.27	1.39	34.8	26.8	15.6	2.330	0.92	6.7
	A12	15-28	4.6	0.20	0.79	0.48	0.52	88.2	42.3	24.6	1.920	1.17	12.8
	IIA13	28-43	4.6	0.13	0.25	0.15	0.10	55.4	23.4	13.6	1.110	1.32	12.3
	IIB2	43-56	4.8	0.05	0.09	0.05	0.02	39.8	7.4	4.3	0.256	3.20	16.9
	IICox	56-82	5.2	0.05	0.12	0.08	0.02	26.4	2.0	1.2	0.089	1.27	12.9
	IICn	82+	5.3	0.10	0.25	0.15	0.02	25.0	0.6	0.3	0.025	1.15	13.2
TV23	A11	0-18	4.7	0.09	0.93	25.99	6.85	40.7	14.7	8.5	0.878	2.23	9.7
	IIA12	18-38	4.9	0.05	0.37	2.00	0.32	33.6	8.7	5.1	0.503	2.96	10.0
	IIB21	38-91	4.6	0.05	0.20	0.72	0.08	26.4	5.3	3.1	0.288	2.70	10.7
	IIB22	91-104	4.8	0.10	0.22	1.04	0.09	23.4	1.7	1.0	0.101	2.96	9.7
	IIC1ox	104-168	5.2	0.05	0.08	0.59	0.04	16.6	1.1	0.7	0.040	2.47	16.5
	IIC2ox	168-176	5.5	0.05	0.05	0.53	0.03	17.3	0.7	0.4	0.028	3.70	14.3
	Ab	176-186	5.3	0.10	0.09	1.04	0.07	25.0	0.1	<0.1	0.010	2.47	7.0
	B21b	186-199	5.0	0.10	0.19	1.20	0.09	27.8	0.1	<0.1	0.009	2.96	7.8
	IIB22b	199-217	5.3	0.10	0.20	1.04	0.09	33.9	0.3	0.2	0.011	3.20	15.5
	IIB23b	217-247	5.2	0.10	0.15	1.04	0.09	23.4	0.2	0.1	0.010	2.96	10.0
	IIC1oxb	247-277	5.1	0.05	0.25	0.98	0.09	30.2	0.4	0.2	0.003	2.96	66.7
	IIC2oxb	277+	5.0	0.05	0.20	0.95	0.10	26.4	0.1	<0.01	0.003	2.84	23.3

* Section sites are on figure 2.

rare polytype of kaolinite with weak reflections at 3.58 Å (004) (Carroll 1970), in the buried paleosol, and its absence in the relict paleosol provides an important means of differentiation. Small amounts of quartz, feldspar, and mica in the buried paleosol suggest intense weathering over a long interval of time. The relatively low amount of clay minerals present may reflect a lower Si content in Mt Kenya volcanic rocks, and a high rate of silica removal by leaching under a humid tropical mountain climate. In the A11 horizon relatively high amounts of quartz, feldspar, and mica are suggestive of aeolian influx. The absence of gibbsite may result from its incorporation into clay mineral lattices.

No attempt was made to date the Ab horizon in the buried paleosol as it is considered beyond the range of radiocarbon. Moreover, organic matter and organic carbon distributions (table 3) in the overlying relict profile suggest the possibility of contamination by leaching, which would make dating difficult. Dithionite-extractable iron oxide is slightly higher in the buried soil relative to the overlying relict soil. On the whole one might expect even greater values for free Fe_2O_3, given the oxidizing potential of a strongly acid to very strongly acidic environment in the buried soil, and the apparent differences in age indicated by contrasting particle size and clay mineral distributions. The uniform pH in the buried soil indicates little leaching at depth; however, the increase in pH in the relict soil indicates moderate leaching which is surprising given the high rainfall, luxuriant bamboo forest, and low quantities of common 2:1 clay minerals. In addition, the CEC's indicate sufficient cation exchange capacity to support larger amounts of illite and kaolinite. High Ca^{+2} in the A11 horizon may be due to the combination of aeolian influx and base recycling by plants, a phenomenon observed in the subalpine forests of western North America (Mahaney 1973, 1974; Mahaney & Fahey 1976).

TILLS WITH WELL PRESERVED MORAINAL FORM

End moraines deposited during the last glacial maximum (Liki I) are found as low as 3 100 m, although ground moraine is found as high as 3 600 m. These deposits carry significantly different weathering features indicating a younger age. Boulder frequency is higher than on the surface of Teleki till. Weathering ratios indicate substantial differences from older drift, and weathering rinds are considerably thinner. The type locality in Teleki valley (figure 2) carries a soil with the following description:

TV22: post-Liki-I soil (location figure 2)

Horizon	Depth (cm)	Description*
01	4-0	Black (10YR 2/1m).
A11	0-15	Black (10YR 1.7/1m) silty clay loam, granular structure, friable, plastic and non-sticky.

*Colors are from Oyama and Takehara (1970) and are given as moist (m) and dry (d).

33

Horizon	Depth (cm)	Description*
A12	15-28	Brownish black (10YR 2/2m) silt loam, weak granular structure, friable, plastic and slightly sticky.
IIA13	28-43	Black (10YR 2/1m) and dark brown (10YR 3/3d) silty clay loam, weak granular structure, friable, plastic and slightly sticky.
IIB2	43-56	Dark brown (7.5YR 3/4m) and dull brown (10YR 5/4d) pebbly loam, weak blocky structure, firm, plastic and sticky.
IICox	56-82	Brown (10YR 4/4m) and dull yellow orange (10YR 6/4d) pebbly sandy loam, massive structure, firm, plastic and sticky.
IICn	82+	Light gray (2.5Y 8/1m), pale yellow (2.5Y 8/3m) and grayish yellow (2.5Y 7/2d) pebbly loam, massive structure, loose, non-plastic and non-sticky.

*Colors are from Oyama & Takehara (1970) and are given as moist (m) and dry (d).

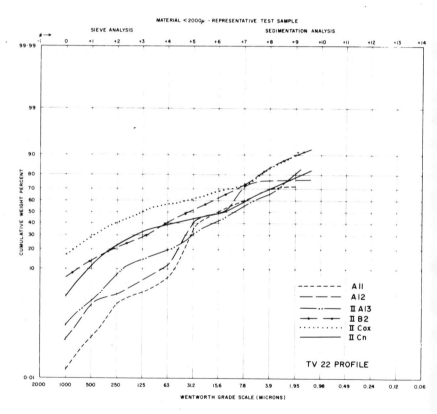

Figure 4. Particle size distributions for soil horizons in TV22 profile formed in Liki-I till.

34

This soil is moderately developed, contains a solum one-half the thickness of the TV23 solum, and a shallow depth. Granulometric data (table 2, figure 4) indicate a lighter texture than in TV23; and the A horizons have significantly higher silt plus clay suggesting an aeolian origin. The frequency distribution center of gravity (mean phi) is 4.8 for the B horizons and 6.7 for the A13. When compared with the granulometry for TV23 these data indicate a shift to the left giving higher sand and lower silt plus clay in the lower solum. A mean phi shift of this magnitude indicates a considerable difference in age.

X-ray reflections indicate that only minor amounts of kaolinite and halloysite are present in the post-Liki-I soil, together with moderate to trace amounts of quartz, feldspar, and mica. Within the primary mineral suites, a major break is seen at the A13/B2 boundary where quartz, feldspar and mica diminish with depth. This may result from aeolian influx and tends to support the particle size data which indicate two parent materials (loess/till, i.e. I/II).

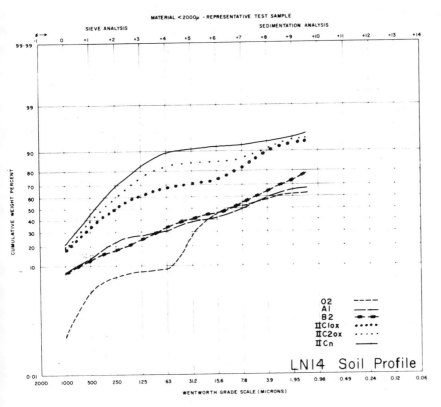

Figure 5. Particle size distributions for soil horizons in LN14 profile formed in Liki-II till.

35

Figure 6. Tyndall Glacier foreland and location of TT2 Type Section in background: proximal slope of the Liki moraine is shown left of center with Teleki Valley Ranger Station on the crest; and valley train of late-Pleistocene/Holocene age in foreground.

There is little indication of leaching in the TV22 profile judging by the increase in pH with depth. Cation exchange capacity (CEC) and extractable cation data suggest that 1:1 and 2:1 clay minerals are possible; their development may be inhibited by low amounts of Si. Organic matter and organic carbon are higher in the A horizon complex suggesting that microbial activity is less intense at timberline in the Hagenia Woodland/Ericaceous Zone transition belt (Coe 1967, Hedberg 1964). This is further substantiated by the presence of an 01 horizon and a higher C/N ratio. Dithionite extractable iron oxide is lower than in the TV23 soil, a factor attributed to less time for development.

Liki-II deposits, located above 3 800 m in most drainages, form conspicuous and well-preserved moraine and outwash systems. These systems are presumed to have formed when ice receded upvalley from its maximum extent during the last pleniglacial (Würm, Wisconsinan). The ice front halted at 4 000 and 4 200 m and built up end and lateral moraines. A short core in outwash associated with these end moraines at site LN5 (figure 2) yielded basal organic matter dated at 12 590 ± 300 BP (GaK-8275). This data provides a minimum age for the Liki-II moraine complex, and a maximum age for the post-Liki-II soil. Cores from four other bogs on the Liki-II moraine complex are presently being analyzed for granulometry, clay mineral content, and radiometric age. Moreover, organic matter buried in lacustrine sediment behind the 4 000 m moraine complex (figure 2) promises to provide at least two additional dates.

Weathering data indicate that boulder frequency is somewhat higher than for Liki-I deposits; weathering ratios are similar; and maximum and minimum rinds are slightly smaller. The data indicate both Liki-I and -II deposits are closely related in age, and distinctly different from older deposits.

The post-Liki-II soil is described as follows:

Horizon	Depth (cm)	Description*
02	0-8	Black (10YR 2/1m) silty clay, weak granular structure, friable, slightly plastic and non-sticky.
A1	0-10	Brownish black (10YR 2/2m) clay, granular structure, firm, plastic and sticky.
B2	10-20	Dull yellowish brown (10YR 5/4m), dark reddish brown (5YR 3/6m) and brown (7.5YR 4/6d) pebbly clay loam, weak blocky structure, firm, plastic and sticky.
IIC1ox	20-33	Dull yellowish brown (10YR 4/3m) pebbly sandy loam, massive structure, loose, non-plastic and slightly sticky.
IIC2ox	33-58	Grayish yellow brown (10YR 4/2m) pebbly loamy sand, massive structure, loose, non-plastic, non-sticky.
IICn	58+	Brownish black (2.5Y 3/2m) pebbly sand, massive structure, loose, non-plastic and non-sticky.

* Colors are from Oyama and Takehara (1970) and are given as moist (m) and dry (d).

The morphological data indicate that a B horizon 10 cm deep formed in 12 590 ± 300 years on the northwest flank of the mountain. The reddish brown (5 YR) hue in this horizon suggests translocation of organic matter

37

which is substantiated by the laboratory data (table 3). The parent material and IIC1ox horizons have coarser textures and less clay, thus reducing plasticity and stickiness. Particle size data (table 2, figure 5) show increasing amounts of sand with depth, and a general fining sequence upward where silt and clay are highest in the solum. This mechanical mix of silt plus clay is considered to result from the combined effects of aeolian influx and pedogenesis. Mean phi values are 6.2 for the B horizon and 6.6 for the A horizon which appear comparable to the post-Liki-I solum.

Clay minerals are largely absent in the LN14 profile and in the TV22 profile and the top 104 cm or the whole solum of the TV23 profile with the exception of a trace occurrence of illite in the IIC1ox horizon. This phenomenon is seen elsewhere in post-Liki-II soils in the upper Afroalpine area (especially at site TV1, Mahaney 1980), and in soils containing basaltic and andesitic clasts on Mt Adam in the Cascadas of W. North America (Mahaney & Fahey, in preparation). Insofar as the primary minerals are concerned quartz

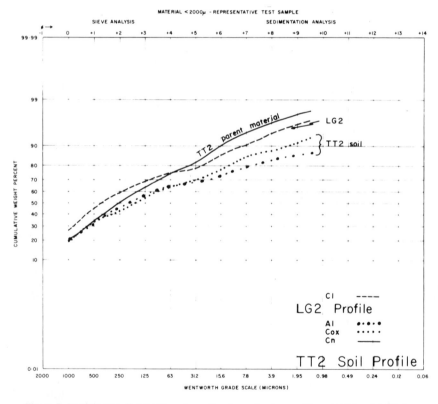

Figure 7. Particle size distributions for soil and weathering profiles at TT2 and LG2 sites.

38

dominates. It gives a pattern which suggests leaching is sufficiently powerful to translocate large quantities from the solum into the subsoil. The high amount of quartz in the 02 horizon may be due to aeolian influx. The distributions of feldspar and mica also suggest some aeolian contribution.

The post Liki-II soil ranges from very strongly acid in the solum to medium acid in the subsoil and parent material. Dominant cations are K^+ and Ca^{+2}, and as with older soils, plant recycling probably accounts for high basic cation content in surface horizons. Organic matter, organic carbon, and nitrogen all diminish with depth suggesting that translocation of organic constituents is small. Free iron is lower than in the post-Liki-I soil, probably a result of younger age and lower acidity.

TILLS IN HIGH VALLEYS

Neoglacial deposits are divisible into two advances on the basis of topographic position, relative weathering features (table 1) and presence/absence of soil development (tables 2 and 3). The oldest deposits belong to the Tyndall advance (figure 2) and are here termed Tyndall till. Boulder frequency is high, stones have unpitted surfaces, and generally lack any surface discoloration. Weathering rinds are small (<3.0 mm maximum rind) and 25 % of all stones lack rinds. The minimum rind is nil. Soil at the type section (TT2; figure 6) is described as follows:

TT2: post-Tyndall soil

Horizon	Depth (cm)	Description*
01	2.5-0	Black (10YR 2/1m).
A1	0-20	Brownish black (10YR 3/2m) and grayish yellow brown (10YR 4/2d) pebbly sandy loam, friable, non-plastic, and non-sticky.
Cox	20-41	Dull yellowish brown (10YR 4/3m, 10YR 5/3d) pebbly sandy loam, friable, non-plastic, and slightly sticky.
Cn	41+	Light gray (2.5Y 7/2m, 7/1d) pebbly loamy sand, loose, non-plastic, and non-sticky.

* Colors are from Oyama & Takehara (1970) and are given as moist (m) and dry (d).

This profile is thin and poorly developed to a depth of 41 cm. Particle size data (table 2, figure 7) confirm the downward movement of clay as suggested by the field data. The particle size curve (figure 7) shows only minor changes in the soil relative to the parent material. Mean phi values amount to 3.1 for the Cox horizon and 3.4 for the A horizon, which are considerably smaller values when compared to older profiles. No clay minerals are present in the profile, but primary minerals such as quartz, feldspar, and mica are all present in minor amounts (table 2). A pH of 5.2 in the surface soil (table 3) indicates that this system is in disequilibrium with the environment (cf. LN14 profile,

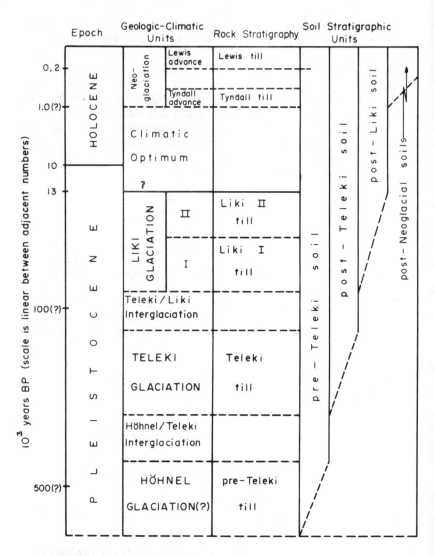

Figure 8. Correlation diagram. Geologic-climatic unit boundaries are based on maximum or minimum dates from non-glacial sediments. Dashes indicate boundaries for which there is no radiometric control. Liki interstades are omitted because there are no data for their definition. Climatic Optimum is considered to be the best term available for a warmer climatic interval following the Liki Glaciation. Neo-glaciation refers to a resurgence of glaciation following the post-Climatic Optimum period.

table 3). Data for extractable cations and CEC show that the soil has a lower total exchange capacity, a condition attributed to insufficient time for development, and lower overall amounts of organic matter. This combined with a smaller amount of nitrogen, produces a lower C/N ratio. According to table 3 free iron oxide in A1 is 0.37 % (as in LG2) and is 0.47 % in the Cox horizon.

The Lewis advance laid down tills which consist largely of an open network of boulders lacking any appreciable soil cover. On the older moraine crests very thin weathering profiles cover perhaps 10 % of the surface area to a depth of approximately 10 cm. Boulder counts are high, weathering ratios give 100 % fresh stones with unpitted surfaces and weathering rinds are < 1.0 mm (table 1). The LG2 profile is described as follows:

LG2: post-Lewis weathering profile

Horizon	Depth (cm)	Description*
C1	0-10	Dull yellow orange (10YR 7/3m; 7/2d) pebbly coarse loamy sand, massive structure, non-plastic and non-sticky.

* Colors are from Oyama & Takehara (1970) and are given as moist (m) and dry (d).

Particle size distribution data (figure 7) correlate closely with the parent material curve for the TT2 profile. A mean phi of 1.9 indicates a sandy texture characteristic of relatively young soils formed in glacial drift (Mahaney 1974, 1975). As with the older deposits of Tyndall age, no clay minerals are present. Somewhat larger amounts of quartz and minor amounts of feldspar appear common, which may indicate lithological differences between the two cirques.

Important differences between the youngest and oldest post-Neoglacial soils can be gleaned from the chemical data in table 3. Soil reaction is close to neutral, and extractable cation data indicate high values for alkaline earths and alkaline metals, presumably the result of fairly rapid hydrolysis in the tropical mountain environment. Cation exchange capacity is low and barely sufficient to support the development of 1:1 clay minerals. Low values for organic matter, organic carbon, nitrogen and C/N ratio indicate the state of the system in the initial stage of development. Free iron oxide is nearly as high as in the TT2 profile (Tyndall age).

CHRONOLOGY

The preceding sections describe the principal glacial units recognized in the Mt Kenya area (figure 8), identify the criteria used in recognition and correlation, and summarize age relationships. Four units are recognized including deposits of pre-Teleki till (undifferentiated), deposits of Teleki till, deposits of Liki till and deposits of Neoglaciation. Older units are exposed near the lower limit of glaciation or in interfluves in the Upper Afroalpine belt; the youngest being in the valley heads. Neoglacial and Liki deposits mantle four-fifths of nearly every glaciated valley, the remaining one-fifth occupied by Teleki and pre-Teleki drifts.Stratigraphic names are taken from type localities along the north and west flanks of the mountain and these are extended to other drainages on the basis of morphology, weathering and soil development.

In the correlation diagram (figure 8) most geologic-climatic unit boundaries are based either on maximum or minimum radiocarbon dates or on relative criteria. In most cases radiometric controls are lacking and most tentative (−) boundaries will likely be shifted down the time column. Work in progress at this time may soon yield absolute ages for Liki-I drift and for a multitude of alluvial fans of Holocene age (Mahaney 1981 in press). Boundaries for pre-Liki-I units are based on correlation with deposits for which radiometric ages have not been determined as yet. Liki interstades are omitted since no evidence for their definition has been found. Climatic Optimum (van Zinderen Bakker & Maley 1979) is considered to be the best term for an interval following the Liki Glaciation when climate was presumably warmer than present. Time boundaries for the Climatic Optimum may be modified subsequently as new dates become available. Evidence from buried soils in alluvial fans built up during the mid-Holocene indicate at least one wet phase between 4 000 and 6 000 radiocarbon years ago (Mahaney 1981). Neoglaciation, and all Neoglacial advances are all informal terms. Tyndall till is a local stratigraphic name for early Neoglacial deposits with indeterminate upper and lower boundaries. Termination of the Lewis advance coincides with glacial retreat documented in other mountainous areas of the world (Mahaney 1974, 1978, 1979, Mahaney & Fahey 1976).

CONCLUSION

Deposits belonging to four major glaciations are differentiated on the basis of radiocarbon ages, topographic position, weathering criteria, and soil development. Representative sections at the type localities are described in detail: changes in horizon sequence, depth of weathering, color, structure, consistence, plasticity, stickiness, clay mineral assemblage, pH, and a variety of soil-chemical parameters are shown to have great utility in deposit differentiation. The greatest immediate need is to produce a large number of bracketing radiocarbon dates for late Pleistocene/Holocene deposits that mantle most major drainages on the mountain. This will allow more positive correlations to be made with adjoining areas, and permit the reconstruction of a detailed time stratigraphy.

ACKNOWLEDGEMENTS

Research was supported by grants from the National Geographic Society and York University. Field work was authorized by the Office of the President,

Geological Survey of Kenya, and Mountain National Parks, Republic of Kenya. I thank L.M.Mahaney, D.Halvorson, B.Blatherwick, R.Blatherwick, L.Gowland and the students in my Mountain Geomorphology course (1976) for assistance in the field. In particular, I am indebted to W.D. and D.Curry, Naro Moru River Lodge, for logistical support, to P.M.Snyder (Assistant Warden) and F.W.Woodley (Warden), Mountain National Parks, Kenya and their rangers for assistance during the course of field work, to M.Bardeck for assistance in the laboratory and to G.Berssenbrugge for preparing maps and diagrams. Bruno Messerli provided helpful comments in the field.

REFERENCES

Baker, B.H. 1967. *Geology of the Mount Kenya Area.* Geological Survey of Kenya, Rep. 79, 78pp.
Carroll, D. 1970. Clay minerals: A guide to their X-ray identification. *Geol. Soc. Amer. Spec. Pap.* 126, 80pp.
Coe, M.J. 1967. *The ecology of the alpine zone of Mt Kenya.* The Hague, Junk, 136pp.
Hedberg, O. 1964. Features of Afroalpine plant ecology. *Acta Phytogeographica Suecica,* 49, 144pp.
Mahaney, W.C. 1972. Late Quaternary history of the Mount Kenya Afroalpine area, East Africa. *Palaeoecology of Africa* 6: 139-141.
Mahaney, W.C. 1973. Neoglacial chronology in the Fourth of July Cirque, Central Colorado Front Range. *Bull. Geol. Soc. Amer.* 84: 161-170.
Mahaney, W.C. 1974. Soil stratigraphy and genesis of neoglacial deposits in the Arapaho and Henderson Cirques, Central Colorado Front Range. In: W.C.Mahaney (ed.), *Quaternary Environments: Proceedings of a Symposium.* Geographical Monographs 5: 197-240.
Mahaney, W.C. 1975. Soils of post-Audubon age, Teton Glacier Area, Wyo. *Arctic and Alpine Res.* 7(2): 141-154.
Mahaney, W.C. 1978. Late-Quaternary stratigraphy and soils in the Wind River Mountains, Western Wyoming. In: W.C.Mahaney (ed.), *Quaternary Soils* Norwich, Geoabstracts: 223-264.
Mahaney, W.C. 1979. Quaternary stratigraphy of Mt Kenya: a reconnaissance. *Palaeoecology of Africa* 11: 163-170.
Mahaney, W.C. 1980. Late Quaternary rock glaciers, Mount Kenya, East Africa. *J. Glaciology* 25(93).
Mahaney, W.C. 1981. Paleoclimate reconstructed from paleosols: evidence from the Rocky Mountains and East Africa. In: W.C.Mahaney (ed.), *Quaternary Paleoclimate.* Norwich, Geoabstracts (in press).
Mahaney, W.C. & B.D.Fahey 1976. Quaternary soil stratigraphy of the Front Range, Colorado. In: W.C.Mahaney (ed.), *Quaternary stratigraphy of North America.* Stroudsburg, Pa., Dowden, Hutchinson & Ross: 319-352.
Oyama, M. & H.Takehara 1970. *Standard soil color charts.* Japan Research Council for Agriculture, Forestry and Fisheries.
Van Zinderen Bakker, E.M. & J. Maley 1979. Late Quaternary palaeoenvironments of the Sahara region. *Palaeoecology of Africa* 11: 83-104.
Zeuner, F.E. 1949. Frost soils on Mount Kenya and the relation of frost soils to aeolian deposits. *J. Soil Science* 1: 20-30.

A 5000-YEAR OLD POLLEN SEQUENCE FROM SPRING DEPOSITS IN THE BUSHVELD AT THE NORTH OF THE SOUTPANSBERG, SOUTH AFRICA

L. SCOTT

Institute for Environmental Sciences, University of the OFS, Bloemfontein, South Africa

ABSTRACT

Analysis of pollen from spring deposits in the Northern Transvaal shows that bushveld vegetation generally not much different from that of the present existed north of the Soutpansberg during the last 5,000 years. A slight change in the vegetation to an apparently more open type of woodland is suggested between about 3,000 and 2,000 years BP.

INTRODUCTION

In South Africa, Quaternary pollen analysis is to a large degree limited by the scarcity of suitable deposits. Most of the work in the relatively dry interior of the country has been done on rare peat or organic deposits associated with isolated springs like at Florisbad (Van Zinderen Bakker 1957), Aliwal North (Coetzee 1967), Alexandersfontein (Scott 1976), Wonderkrater (Scott & Vogel 1978, Scott, a, in preparation) and Rietvlei (Scott & Vogel, in preparation) (Figure 1). Suitable lake deposits are absent and although studies on cave deposits are now in progress, they generally contain very little pollen and in the past have been of limited value. For palynology spring or swamp deposits have their own inherent problems such as the over-representation of pollen produced by the local vegetation and the possibility of fresh roots or other contaminations which may affect the ^{14}C-age determinations (Scott, a). However, spring material remains a good source of information and the present study of the site on the farm 'Scot' in the northern Transvaal provides evidence on late Holocene vegetation and climate in the bushveld bordering the semi-arid Limpopo valley area.

THE SITE

The organic deposit of the farm 'Scot', near Vivo, is situated at about 22°57'S, 29°24'E in the lowveld of Northern Transvaal (Figure 2). The farm lies at the

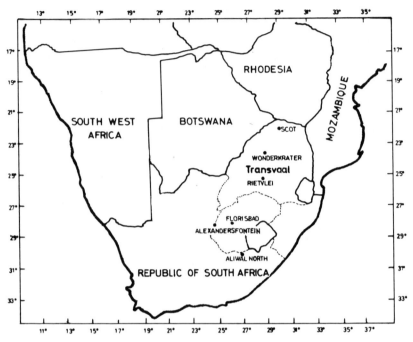

Figure 1. Locality map.

southern limit of the dry Limpopo valley which is bordered by the northern
slopes of the Soutpansberg. The site is situated at about 823 m altitude against
the foot of the mountain which rises to more than 1500 m above sea-level
from the plains which gradually descend to the north over about 80 km to
almost 400 m altitude at the Limpopo River.

The organic material accumulated in a swampy basin ('vlei', Figure 3) with
sands around a strong spring originating from a fault which separates the
Waterberg sandstones of the Soutpansberg from the Karoo basalts of the
plains to the north. More springs are found along this fault for instance those
adjacent to the large saltpan about 6 km to the west of the site.

The climate of the area belongs to the warm and relatively dry zone of the
low-lying Limpopo valley. The bordering Soutpansberg range has slightly less
severe conditions. The area receives virtually no frost in winter and the sum-
mers are very hot with mean daily maximum temperatures in January exceed-
ing 30°C (Schulze & McGee 1978). According to the map of the thermal and
moisture regions of Southern Africa (Poynton 1971), which is based on the
Thornthwaite classification system, the climate can be classified as 'warmer,
temperate'. Rains occur in summer mainly under influence of the Inter-
Tropical Convergence Zone (ITCZ) and the annual precipitation amounts to

46

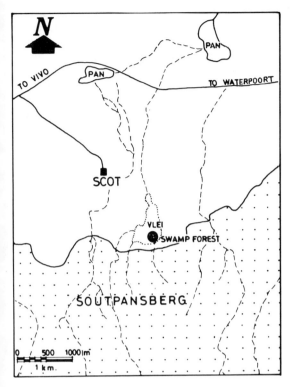

Figure 2. Map showing the position of the swamp forest on the farm Scot.

roughly 400 mm. The site belongs to a narrow zone of so-called 'dry sub-humid' climate between the 'moist sub-humid' and 'semi-arid' zones of the mountain range and the Limpopo valley respectively (Poynton 1971).

THE VEGETATION AROUND THE SITE

The spring and vlei system at Scot, where the peat accumulated, is surrounded by two types of bushveld. To the north lies the dry savanna of the plains which belongs to the Arid Sweet Bushveld (14) (Acocks 1953, veld type number in brackets) to the south the Dry Mountain Bushveld (Werger & Coetzee 1978) against the slopes of the mountain. According to the map of Acocks (1953) the Arid Sweet Bushveld (14) changes into Mopani Veld (15) towards the north and the east, while the area to the south and west is, except for the Sourish Mixed (19) and Sour Bushveld (20) on the mountains and slopes, largely covered by Mixed Bushveld (18). Towards the east the Sour Bushveld of the mountain changes to the wetter NE Mountain Sourveld (8). The

47

relationships between these types and the others of the Transvaal and surrounding areas are discussed by Werger & Coetzee (1978) and Scott (a, in preparation).

The Arid Sweet Bushveld (14) of the plains directly surrounding the spring site occurs on sandy soils and has strong affinities with the *Terminalia* Sandveld vegetation (Werger & Coetzee 1978). A wide variety of trees and shrubs of this bushveld were recorded around the site. They include Combretaceae, especially *Terminalia sericea,* and also acacias, giant euphorbias, Capparaceae, *Rhus* species and many others. Characteristic tree species of the Dry Mountain bushveld (Werger & Coetzee 1978), like *Adansonia digitata, Commiphora* ssp. and a variety of others as well as shrubs like *Myrothamus flabellifolius* were recorded on the rocky slopes south of the spring site (Figure 3). According to the local inhabitants moist forested ravines with yellowwood trees occur higher up in the mountain.

Around the eye of the spring a dense tall, tropical swamp forest (cf. 'mushitu', Van Zinderen Bakker 1969) flourishes (Figures 3 and 4). It contains mainly extra-zonal (azonal) components typical of gallery forests which occur along rivers in the dry tropical bushveld of the lowveld area. The size of the forest is about 20000 m^2 and contains many woody species of which the following were recorded: *Albizzia versicolor, Anthocleista grandiflora, Artobotrys brachypetalus, Berchemia zeyheri, Cassia petersiana, Euclea natalensis, Ficus burkei, F.capensis, Grewia* cf. *hexamita, Maesa lanceolata, Pereskia aculeata, Psoralea pinnata, Rauvolfia caffra, Securinega virosa, Syzygium guineense, Trema orientalis.* There are also some intruders from the surrounding savanna like *Acacia* cf. *nilotica, A.* cf. *tortilis, Dichrostachys cinerea, Rhus leptodictya* and *Terminalia sericea.* On the northern side of the forest many tall trees of *Eucalyptus* sp. have been introduced (Figure 3). Some grasses were recorded around the fringe of the forest while herbs like *Ageratum conyzoides, Hibiscus surattensis, Momordica balsamina, Pycreus polystachyos* and *Senecio pterophorus,* occur on the sparsely vegetated forest floor.

The limited flow of the spring water to the north of the forest passes through the open swampy vlei area (cf. 'dambo', Walter 1964, Van Zinderen Bakker 1969) of about 0,3 km^2 and it terminates in a small enclosed pan about 3 km to the north. The vegetation of the vlei consists mainly of grasses and sedges.

METHODS

The sediment samples were obtained by means of a Hiller peat auger. The palynomorphs were extracted from the peat and organic samples by boiling them in 10 % KOH for 5 minutes, followed by mineral separation with ZnCl$_2$ solution (SG ±2). The pollen was mounted in glycerine jelly and the analyses were made using a Zeiss photomicroscope and a 100 x oil-immersion objective.

Figure 3. View of the swamp forest at Scot from the dry northern slope of the Soutpansberg, showing the open vlei area and the bushveld plains in the distance.

Figure 4. A large specimen of *Ficus burkei* in the swamp forest at Scot.

The analyses of the samples consist of a first count of 200 palynomorphs including the whole spectrum which largely represents local elements and a supplementary count of 200 of the scarcer types from the wider surroundings in order to provide more accurate percentages of the regional pollen. The latter mainly represents the pollen sum (below) which is used for interpretations of the regional vegetation.

THE DEPOSIT AND ITS AGE

The organic deposits filling the swampy area accumulated under moist conditions created by the spring. The sediments of about 1 m depth overlie pinkish sands. It is not certain if there are any older peaty deposits below the sands. Two borings were made by E.M.van Zinderen Bakker Sr in the swamp forest. The 1 m core from Borehole 1 comes from close to the spring eye, and the 0.8 m one from Borehole 2 is from the eastern margin of the forest. Because of its high sand content the sediment cannot be described as peat. It consists largely of fine to medium sand saturated with black peat. Visually no distinct stratification is noticeable in either of the two cores. The following four radiocarbon dates for the cores were provided by J.C.Vogel:

Borehole 1:	−20 to −35 cm:	2460 ± 50 yr BP (Pta-1391, 4502);
	−80 to −100 cm:	3530 ± 60 yr BP (Pta-1392, 4484);
Borehole 2:	−60 to −70 cm:	5070 ± 60 yr BP (Pta-1388, 4503);
	−70 to −80 cm:	4930 ± 70 yr BP (Pta-1387, 4510).

The contradictory dates from two adjacent samples of Borehole 2 can be attributed to inaccurate measurements due to low carbon content of the samples especially in Pta-1387 (J.C.Vogel, personal communication).

THE POLLEN DIAGRAMS WITH DESCRIPTIONS AND INTERPRETATIONS OF POLLEN ASSEMBLAGES − ZONES

The pollen diagrams (Figures 5 and 6) are constructed on the basis of the distinction between the so-called pollen sum (Faegri & Iversen 1964) and the local swamp, grass and other elements. The pollen sum consists of the regional pollen of the wider surroundings presented together as 100 % while the local elements are shown as percentages of the total spectra. The distinction between these groups is arbitrary because it is not always possible to separate the local and regional pollen. The sum is divided into different groups of arboreal pollen or AP (A, B, C and D) and non-arboreal pollen or NAP (E and F) (Figure 5). The AP excludes the tree pollen of the local swamp forest (G).

A photographic record of the fossil taxa from the Transvaal is provided by Scott (b, in preparation) together with assessments of their indicator

49

Table 1. Percentages of some important taxa recorded in surface pollen samples

Altitude (m)	Vegetation type	AP											
		Podocarpus	Burkea africana	Canthium	Proteaceae	Oleaceae	Euphorbia	Combretaceae-type	Rhus	Grewia	Pseudolachnostylis-type	Euclea	Spirostachys
1250 to 1800	Upland Bushveld (5 samples)	<0.6	<3.5		1 to 22	<1	<2.5	<15.8	<15	<0.5		<3.2	<
1100 to 1190	Broad-ortho-phyll Plains Bushveld (5 samples)		<4		<1	<0.5	<0.5.	0.2 to 46		<4	<0.5	1.8 to 15.2	<
1100	Dry Mountain Bushveld (1 sample)		1.5	14		1	0.4	14.4		0.4	7	3.4	
700 to 800	Arid Sweet Bushveld (2 samples)		3.5	0.4	<1			16.5 to 49	<1.5	<1		4 to 11.5	<
832	Swamp forest (1 sample)	0.1	0.5			1	8	1.5	3,5		1.6	2.4	(

values as fossils. Brief mention is also made of aspects such as the distribution and pollen dispersal characteristics of their parent plants. Interpretations of pollen zones are based on these assessments as well as on surface pollen studies of different vegetation types in the Transvaal (Scott, a, in preparation). The latter investigation (Table 1) shows the following:

a) Surface samples from the Arid Sweet Bushveld plains surrounding the site and the area further to the north between Vivo and Alldays contain up to nearly 50 % Combretaceae pollen probably largely derived from the common *Terminalea sericea.*

b) A surface sample from the Dry Mountain Bushveld of the northern slopes of the Soutpansberg directly south of the site differs in containing ±15 % Combretaceae, 15 % *Canthium,* 8 % *Euphorbia* and smaller numbers of others in the AP.

c) The sample from the swamp forest (shown in Table 1 as well as Figure

| | | NAP | | | | | | | | Swamp-forest | | | | | |
Mimosoideae (+ Acacia)	Peltophorum africanum	Capparaceae	Aloe-type	Acanthaceae	Compositae (various)	Stoebe	Anthospermum	Aizoaceae	Cheno-/ Amaranthaceae	Anthocleista grandiflora	Trema orientalis	Rauvolfia caffra	Myrtaceae	Cyperaceae	Gramineae
6 <4		<1	<1	<3	1 to 31	<1.5	<9.5	<0.6	<1					<9.5	30 to 55.5
5 <6		<0.6	<1.8	<2	<4	<10	<2.5	<0.5	<0.5	<5.8				<6.8	8 to 70
2 0.8			1.6	1			3	1	1.4					2	33
<5 <1.6			<6.5	<3.5	<2	<0.5	<1.8	<4	<2				<0.4	<2.5	12 to 33
		0.1	0.2	0.8	2	0.2	0.2	1	0.4	12.5	1	1.8	10	1	48

5, but adapted for the pollen sum in the latter) contains most of the above-mentioned elements but includes pollen grains of the forest. Myrtaceae is the most prominent of these but represents the *Syzygium* trees at the forest as well as the introduced *Eucalyptus* plantation.

d) Five surface samples from distant woodland types show that in general the Broad-orthophyll Plains Bushveld spectra are not very different from those of the Arid Sweet Bushveld, but contain up to about 4 % *Burkea* and 15 % *Euclea* pollen and smaller numbers of others in the AP with especially Compositae (<10 %), Chenopodiaceae-type (<5.8 %), Acanthaceae (<4) and others in the NAP. Generally high percentages of Gramineae (9-80 %) occur in these spectra.

e) Surface samples from the upland woodland types (Sourveld (20) and NE Mountain Sourveld (8)) show fairly high numbers of Proteaceae pollen (up to 22 %) but much smaller percentages of other bushveld tree pollen so that in

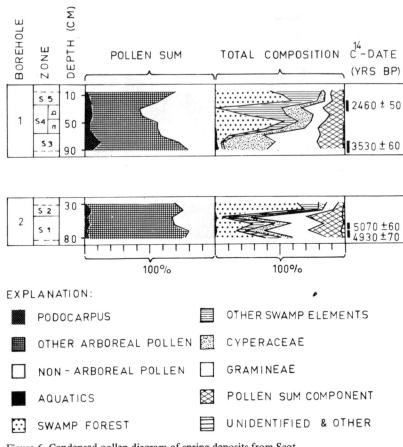

Figure 6. Condensed pollen diagram of spring deposits from Scot.

total, the AP is considerably smaller than the ± 50 % or more as recorded in the warmer, dry, low-lying types. It is also significant that the numbers of Compositae pollen grains are comparatively high (up to 31 %) in the upland woodland types.

The pollen diagrams of the cores from 'Scot' are presented in detail in Figure 5 (in pocket) and summarised in Figure 6. Five pollen assemblage zones with two sub-zones can be distinguished in the two cores from which interpretations of regional and local vegetation changes are derived. The successive designation of these zones (S1 to S5) is adopted for convenience and is not intended to show that Borehole 1 follows upon Borehole 2 stratigraphically. The associated palaeoclimatic conditions are estimated in the case of the reconstructed vegetation types by means of comparisons with surface pollen samples and the present vegetation.

Borehole 2. The core with the bottom-date of about 5000 BP between 60 and 80 cm is divided into two zones, S1 and S2. The AP of the lower zone S1 consists of high numbers of Combretaceae (up to ±55 %) and smaller percentages of various other tree pollen types. This combination suggests that a savanna vegetation occurred in the area shortly after 5000 BP. The NAP consists of mainly Aizoaceae-type (up to ±25 %), some Compositae pollen (up to ±10 %) and smaller numbers of others which indicate fairly dry conditions. Among the local pollen types relatively low numbers of swamp forest elements like Myrtaceae (up to ±28 %) occur in comparison with the rest of the sequence. Swamp elements like Cyperaceae (up to ±15 %) are, however, higher than in the upper zone. The pollen of grasses is also prominent (up to ±45 %) but it is not known what proportions are derived from the local and regional environments.

The upper zone, S2, shows increases in the percentages of *Burkea africana* (up to ±18 %), *Pseudolachnostylis*-type (up to ±11 %) and possibly *Celtis* (up to ±2 %), while Combretaceae decline. In the NAP the Aizoaceae numbers decline while the Compositae show only a slight increase. These changes in the regional vegetation could point to slightly wetter conditions but they are too small to be conclusive. Among the local pollen grains the swamp forest trees, *Trema orientalis* (up to ±6 %), *Rauvolfia caffra* (up to 5 %) and especially Myrtaceae (up to 60 %+) show a sharp increase. This suggests that the 'vlei' area which had previously been as open as at present became overgrown by swamp forest. Whether this is the result of local succession or slightly more humid conditions is uncertain.

Borehole 1. Borehole 1 with the dates of about 3530 BP at 80 to 100 cm and 2460 BP at 20 to 30 cm, is divided into three zones S3, S4 and S5. The middle one is subdivided into two sub-zones S4a and S4b which reflect smaller changes in the zone. In general the composition in S3 is similar to S1 described above and the change from S3 to S4 in Borehole 1 is in many ways comparable to the change from S1 to S2 in Borehole 2, which suggests possible partial overlapping between the cores. This possibility is supported by the available ^{14}C dates. In zone S4 there is, however, a clear decline in the total AP percentages which already started in S3 (Figure 6). The lowest AP:NAP-ratio is reached in sub-zone S4b when the Compositae grains attain their highest percentages of the pollen curve (up to ±53 %). Comparable spectra developed at Wonderkrater in the central Transvaal bushveld (Figure 1) at roughly the same time, viz. between 4000 and 2000 BP (Scott, a, in preparation). These more open assemblages resemble those of surface pollen samples from the Upland Bushveld types (Table 1) and possibly point to slightly cooler conditions. During the interval of sub-zone S4b the swamp forest elements are also largely replaced by monolete fern spores. This local event can possibly be attributed

53

to factors such as plant succession related to increased spring activity or increased precipitation whereby the swamp forest was for a while replaced by open water and ferns. Alternatively, if the slightly cooler climate, as tentatively suggested by the pollen sum elements above, really developed, the forest could have been reduced by such conditions which in the upland regions towards the south are characterised by the greater expectancy of frost (Schulze 1965:132, 136).

In the upper zone S5 the AP of the pollen sum, however, increases and shows that the bushveld again became denser after 2460 BP and that it contained higher proportions of *Burkea africana*. At the same time the swamp forest types were restored so that it became more extensive than at present. Trilete fern spores are also prominent in zone S5 while the Cyperaceae and Gramineae pollen grains remain relatively unimportant. In general this zone seems to suggest slightly wetter conditions than at present.

DISCUSSION

The interpretation of the pollen diagrams suggests that bushveld vegetation occurred in the area during the last 5000 years but that some changes in its composition took place. Although other factors such as the possible shifting of the spring-eye and the associated swamp forest, or the clearance of the forest or bush by man, can at this stage not be excluded with definite certainty as possible causes for the changing tree cover, it is not unlikely that these oscillations were caused by small climatic changes as suggested. In connection with possible human influence, however, pollen evidence for maize cultivation at Wonderkrater in the central Transvaal plains a few hundred years ago (Scott, a, in preparation) did not seem to coincide with noticeable destruction of the tree-cover. Man will, therefore, not necessarily have altered the situation to such an extent that it can be recorded in the pollen diagrams.

The cause and mechanism of relatively small scale climatic oscillations in the Holocene in South Africa is problematical and future explanations of the processes and changing controlling factors involved, depend on more palaeoclimatic data.

ACKNOWLEDGEMENTS

I should like to thank Prof. E.M.van Zinderen Bakker for providing me with the peat cores from Scot which he sampled in 1971. I also want to thank Dr J.C.Vogel of the CSIR, Pretoria who is responsible for the ^{14}C age determinations. Mrs E.van Hoepen of the Botanical Research Institute in Pretoria is thanked for the identification of plant material. Mr W.J.Helm of the farm 'Scot' kindly allowed me to work on his premises.

REFERENCES

Acocks, J.P.H. 1953. Veld types of South Africa. *Mem. Bot. Surv. S.Afr.* 28:1-192.

Coetzee, J.A. 1967. Pollen analytical studies in East and Southern Africa. *Palaeoecology of Africa* 3:1-146.

Faegri, K. & J.Iversen 1964. *Textbook of pollen analysis* Blackwell, Oxford, 237pp.

Poynton, R.J. 1971. A silvicultural map of Southern Africa. *S.Afr. J. Sci.* 67:58-60.

Schulze, B.R. 1965. Climate of South Africa. *S.A. Weather Bur.* 28. The Government Printer, Pretoria.

Schulze, R.E. & O.S.McGee 1978. Climatic indices and classifications in relation to the biogeography of southern Africa. In: M.J.A.Werger (ed.), *Biogeography and ecology of southern Africa.* Junk, The Hague:19-52.

Scott, L. 1976. Preliminary palynological results from the Alexandersfontein Basin near Kimberley. *Ann. S.Afr. Mus.* 71:173-189.

Scott, L. a A late Quaternary pollen record from the Transvaal bushveld, South Africa. *Quat. Res.*

Scott, L. b. Late Quaternary fossil pollen grains from the Transvaal, South Africa. *Rev.Palaeobot. Palynol. 3b.*

Scott, L. & J.C.Vogel 1978. Pollen alaysis of the thermal spring deposit at Wonderkrater (Transvaal, South Africa). *Palaeoecology of Africa* 10:155-162.

Scott, L. & J.C.Vogel. A Holocene pollen profile from the southern Transvaal highveld, South Africa, in preparation.

Van Zinderen Bakker, E.M. Sr. 1957. A pollen analytical investigation of the Florisbad deposits (South Africa). In: J.D.Clark (ed.), *Proc. 3rd Pan-Afr. Congr. Prehist. Livingstone (1955).* London, Chato & Windus:56-67.

Van Zinderen Bakker, E.M. Sr. 1969. The Pleistocene vegetation and climate of the Basin. In: J.D.Clark (ed.), *Kalambo Falls Prehistoric site 1.* Cambridge Univ. Press: 57-84.

Walter, H. 1964. *Die Vegetation der Erde 1.* Gustav Fisher Verlag, Jena, 592pp.

Werger, M.J.A. & B.J.Coetzee 1978. The Sudano-Zambezian Region. In: M.J.A.Werger (ed.), *Biogeography and ecology of southern Africa.* Junk, The Hague:301-462.

NEW EVIDENCE FOR A LATE PLEISTOCENE WET PHASE IN NORTHERN INTERTROPICAL AFRICA

R. ALAN PERROTT
School of Biological and Environmental Sciences, New University of Ulster, Coleraine, N.Ireland

F. A. STREET-PERROTT
School of Geography, Mansfield Road, Oxford, United Kingdom

ABSTRACT

Workers in a number of closed lake basins in tropical Africa have described lacustrine formations which have been cited as evidence for a major wet period between 30 000 and 20 000 BP. The validity of this phase has always been open to question, because the chronological framework relies too heavily on carbonate dates. This paper presents new stratigraphic and palynological evidence from the Aberdare Plateau, Kenya, indicating that there was an important forest period prior to the last glacial maximum, which culminated after 24 000 BP. A review of the most reliable lacustrine chronologies suggests that the greatest extent of lakes occurred between 25 000 and 22 000 BP. This wet phase was most marked between the equator and about 22°N. It was not experienced in the equatorial lakes, but is clearly evident in deep sea cores which record long-term fluctuations in the discharge of the Niger and the Nile. We attribute the lake expansion to a combination of lower temperatures, slightly increased precipitation and possibly increased cloud cover and runoff.

INTRODUCTION

Nearly 20 years ago, the advent of radiocarbon dating initiated a new era in Quaternary research on the African continent. In one of the first studies to apply the method to lacustrine sediments, Faure *et al.* (1963) established that the last major phase of expanded lakes in the southern Sahara had occurred during the early and mid-Holocene. In their list of ^{14}C dates they reported a single measured age of 21 350 ± 350 BP on lake marl with *Phragmites* from the Fachi depression (18°N), a small independent drainage basin within the Chad catchment. At the time, it was impossible to interpret this isolated date (Faure 1969). But subsequent stratigraphic work by Michel Servant indicated that a major lacustral episode had occurred in the Chad Basin between 30 000 and 20 000 BP (Servant 1973). The existence of this Late Pleistocene wet phase was supported by a number of later studies in eastern Africa and the southern Sahara; the most detailed chronology being derived from Lake Abhé (11°N) on the border between Ethiopia and Djibouti (Gasse & Delibrias 1977). Not all investigators were convinced by the lacustrine evidence, however. The eminent stratigrapher Professor Karl Butzer has consistently questioned the occurrence of a lacustral episode between 30 000 and 20 000 BP (Butzer *et al.* 1972, Butzer 1979, personal communication). His telling criticisms of the published data have three main grounds: 1) the lack of any evidence for an expansion of Lake Turkana (5°N) during the interval 35 000-9 500 BP; 2) the dependence of most studies on carbonate dates, which become increasingly unreliable before 20 000 BP; and 3) the lack of supporting evidence from other indicators such as pollen and fluvial sequences.

In this paper, we first present new pollen evidence from the Aberdare Plateau in Kenya (0°S) and then re-examine the general stratigraphic and dating framework for the period 30 000-20 000 BP in the light of Professor Butzer's criticisms. The upper Late Pleistocene wet phase, which is now dated in the Aberdares by ^{14}C measurements on organic mud, also features in recently published chronologies derived from organic matter in marine cores with high sedimentation rates (Stanley & Maldonado 1977, 1979; Pastouret *et al.* 1978). Finally, we attempt to summarize the limited information on climatic conditions during the period of interest.

NEW POLLEN EVIDENCE FROM THE ABERDARE PLATEAU

The Aberdare Range (0°10'-0°45'S) runs from north to south on the eastern flank of the Kenyan Rift Valley. It is a major catchment area for streams flowing into the Rift-floor lakes of Naivasha and Elmenteita, and as such, climatic conditions on the Range have a considerable influence on water levels in the lakes.

Location description

The Aberdare Plateau is an extensive area of moorland with dense ericaceous thickets and large sedge-dominated mires on the lower lying areas (ca.3 000 m). *Alchemilla*-dominated grassland flanked by open *Hagenia-Hypericum* woodland occupy the higher ground and slopes.

A section of the southern end of the very extensive Karimu Mire (0°30'S, 36°41'E) at an altitude of 3 040 m was selected for study. The area investigated was a south-eastern embayment of the mire, bounded to the north-east by a ridge of higher ground with dense ericaceous thickets and on the east by a low *Alchemilla*/Gramineae-covered rise through which the excess waters of the mire outflow in periods of higher precipitation. On the southern flank, the mire extends up over three gradually rising terraces to meet a steep boundary slope with open *Hagenia-Hypericum* woodland.

A transect of 16 borings was taken across the mire and terraces by means of a 'Hiller' corer and sample pits to establish the stratigraphic relationship of the underlying deposits. The results are illustrated in a schematic form in figure 1. Samples from points K, N and P were submitted for radiocarbon determination.

The uppermost deposit consists of peat with ranges in depth from ca.50 cm on the upper terraces to ca.250 cm on the main part of the mire. Three distinct variations in peat type are encountered; an upper loose, wet, dark-brown peat, becoming firmer with depth, which grades down into a firm, red-brown peat overlying a grey-brown peat with a considerable inorganic content. The only radiocarbon determination so far available from the peat horizon was taken from just above the boundary between the red-brown and grey-brown peats and yielded the following result:

Point K — 210-230 cm — SRR-1123 — $9\,504 \, ^{+\,50}_{-\,50}$ BP.

This peat sequence extends across the main part of the mire and onto the first terrace to ca.40 m beyond Point P. Only the dark brown peat continues across the upper terraces to the mire edge.

On the mire and the first terrace, the base of the peat forms a sharp boundary with a light coloured inorganic clay ranging in depth from 20 cm at point P to $\leqslant 350$ cm on the main part of the mire. No trace of this clay horizon is found on the second and third terraces.

Underlying the clay on the mire and first terrace, and the dark brown peat on the upper terraces, is a stiff brown organic mud ranging in depth from ca. 50 cm to ca.300 cm, which contains numerous monocotyledonous macrofossils. This mud has been dated in three places (figure 1) with the following results:

Point P — 280-300 cm — SRR-1126 — $24\,221 \, ^{+\,445}_{-\,425}$ BP

Point N — 410-450 cm — SRR-1125 — $29\,251 \, ^{+\,875}_{-\,790}$ BP

Point K — 520-540 cm — SRR-1124 — $32\,600 \, ^{+\,730}_{-\,670}$ BP

The base of the organic mud grades down into brown inorganic clay which proved difficult to penetrate in places due to its stiff consistency. This appears to be underlain by decaying rock.

Working from the base of the mire upwards, the stratigraphic sequence records a period of undetermined duration prior to 32 600 BP when clay was being deposited in the basin. The absence of organic material from the clays would seem to imply that vegetation was sparse or absent from the surrounding area. Shortly before 32 600 BP, conditions began to ameliorate and a more organic deposit began to form in the lowest part of the basin. This would appear to have been a period of climatic instability, since the formation of an organic mud/peat alternated with the deposition of inorganic clay.

Gradually improving conditions and increased moisture led to the development of swampy pools where an organic mud was deposited. If conditions

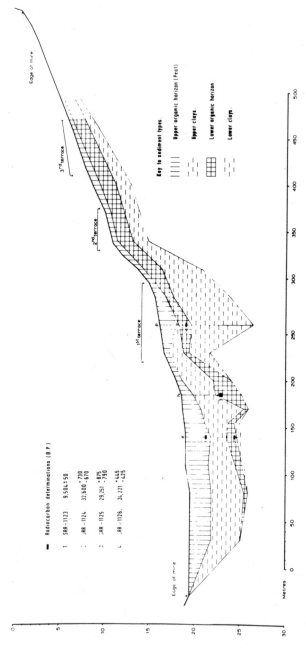

Figure 1. Stratigraphy of the Karimu Mire, Aberdare Range, Kenya.

remained cool, evaporation would have been reduced and it would not have required a large increase in precipitation to allow swamp conditions to exist on the mire. Whatever the climatic interpretation, the presence of organic mud in considerable quantities on the upper terraces would suggest that quite wet conditions prevailed.

This general pattern continued for ca.10 000 years, probably with alternations between open water and sedge-dominated swamp. The organic muds bear a strong resemblance to those deposited in shallow water ($\leqslant 1$ m) on parts of the Karimu Mire today, which suggests that a large body of reed-fringed standing water may have existed on the plateau for at least part of this period.

Some time after 24 200 BP, the trend to aridity was reasserted. Vegetation again became sparse or absent and inorganic clays were deposited on the mire. Dry, cold conditions probably continued until ca.11 000 BP, this being the minimum date for deglacierization of Mt Elgon and the only available date for this transition in Kenya (Hamilton & Perrott 1978, 1979). Peat then began to form in the basin.

Pollen analysis

Pollen analysis has been carried out on cores K and P (figure 1). Pollen diagrams showing selected taxa out of the 72 pollen types identified are presented in figure 2.

In core K, the material at the base of the core contained insufficient pollen to justify analysis, so counts commenced at 500 cm and stopped at 440 cm due to lack of pollen in the clay layer. Virtually no pollen was found in the clays until 275 cm was reached and the counts were recommenced. From the date of 9 504 BP at 210-230 cm we can infer that the 275 cm level represents the base of the Holocene on this mire.

In core P, the quantity of pollen in the sediment decreases rapidly below the base of the organic mud at ca.350 cm. In this case, counts were commenced at 340 cm and continued to the surface; the only gap being between 50 and 85 cm where the samples were lost in transit from Africa. We would, however, suggest that the layer of clay between 188 and 210 cm in core P represents a major discontinuity. It is probable that a greater thickness of inorganic clay was deposited on the terraces during the glacial period but was subsequently eroded during the early Holocene wet phase before initiation of peat growth.

For the purpose of the present study, the core segments we are concerned with are from 500 to 440 cm in core K and between 340 and 210 cm in core P (between the dashed lines in figure 2). We believe that the base of the organic mud at 340 cm in the P core forms a slight overlap with, or is very close in time to, the 440 cm level in core K. We intend to analyze the diagrams as a composite core reading from 500 to 440 cm in core K through 340 to ca.210 cm in core P.

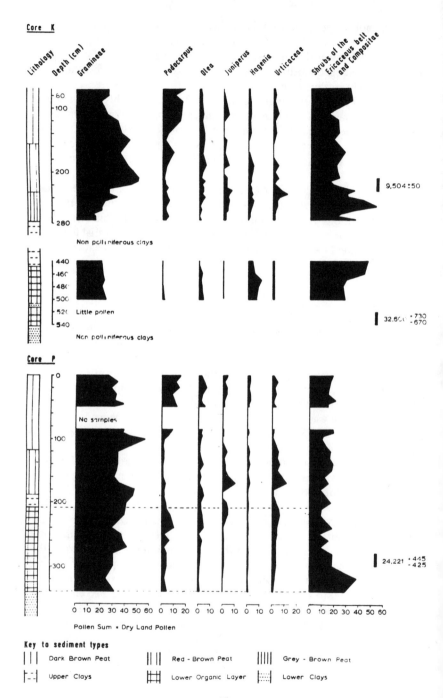

Core K

Lithology Depth (cm) Gramineae Podocarpus Olea Juniperus Hagenia Urticaceae Shrubs of the Ericaceous belt and Compositae

60
100

200

280

Non polliniferous clays

9,504±50

440
460
480
500
520
540

Little pollen

32.600 +730 −670

Non polliniferous clays

Core P

0

No samples

100

200

300

24.221 +445 −425

0 10 20 30 40 50 60   0 10 20   0 10   0 10   0 10   0 10 20   0 10 20 30 40 50 60

Pollen Sum = Dry Land Pollen

Key to sediment types

||| Dark Brown Peat || || Red - Brown Peat ||||| Grey - Brown Peat

|‐-‐| Upper Clays   Lower Organic Layer   Lower Clays

62

If we examine the montane forest pollen at the base of the diagram in terms of the 'relative export ability' of the taxa (Hamilton 1972, Flenley 1973, 1979, Hamilton & Perrott in press), we find that the species of moderate relative export ability, *Hagenia* and *Olea*, re strongly evident, while the species of high relative export ability, *Juniperus, Podocarpus* and Urticaceae, are virtually absent. We suggest that at this time, shortly after 32 600 BP, cold dry conditions had depressed the vegetation belts. Karimu Mire was probably surrounded by a mixed community dominated by Gramineae, *Alchemilla* and Compositae with some Umbelliferae, *Dendrosenecio* and Ericaceae present. Cyperaceae dominated the mire surface. The only tree species occurring near the mire at the time were *Hagenia* and *Olea*. The strength of the *Hagenia* peak (ca.10 %) suggests the existence of open conditions, in which it was regenerating freely, on the Aberdare Range at the time. Today, though an extensive *Hagenia* woodland fringes the southern flank of the mire, the species only represents about 2 % of the pollen sum and never exceeded this percentage during the course of the Holocene. The virtual absence of such well dispersed taxa of the montane forest as *Podocarpus* and *Juniperus* would tend to confirm the hypothesis that the vegetation zones were severely depressed.

The marked expansion of the ericaceous belt species after 470 cm indicates that after an initial improvement climatic conditions again deteriorated.

At the base of the P core, we find evidence that the vegetation zones remained somewhat depressed, though conditions had ameliorated sufficiently to allow the spread of *Podocarpus,* coinciding with a gradual increase in Urticaceae and *Juniperus. Urtica* sp. has been found to be associated with moist conditions on Mt Elgon (Hamilton & Perrott in press) and its increase at this time, together with *Podocarpus* and *Juniperus,* would indicate that increased wetness was responsible for the expansion of montane forest. *Hagenia,* though still present, shows a marked decline from its early strength. It is a light-requiring species and the development of montane forest would restrict the regeneration of *Hagelia* to natural clearings on the forest edge.

This gradual increase in the drier type of montane forest and the decline in the ericaceous belt species continued for some time after 24 200 BP. This latter period bears a strong resemblance to present-day conditions, though it was probably drier and cooler. Conditions were, however, not as severe as prior to 32 000 BP or after ca.21 000-20 000 BP.

COMPARISON WITH OTHER POLLEN DIAGRAMS

The only published pollen diagrams containing detailed records of comparable age are those from Kaisungor Mire on the Cherangani Hills and from Sacred

---

Figure 2. Pollen diagram for the Karimu Mire, showing relative percentages of selected taxa.

Lake on Mt Kenya (Coetzee 1967), with basal dates of 27 750 and 33 350 BP respectively; and from marine cores taken off Senegal (Rossignol-Strick & Duzer 1979a,b), which date back to ca.23 000 BP.

The diagram from Kaisungor Mire records a deterioration in climate just after 27 750 BP which may correlate with the decline in *Hagenia* and *Olea* and the increase in the ericaceous belt species which occurred at ca.465 cm in core K on Karimu Mire.

The Sacred Lake diagram records cold, dry conditions at 33 350 BP when the upper forest limit was severely depressed ($\geqslant$500 m). This was followed by a period of climatic amelioration when forest began to colonize the area around the lake. At about the 13.7 m level, ericaceous belt species increase and *Hagenia* declines, correlating with the climatic deterioration recorded on Karimu and Kaisungor Mires. However, at this site, the *Podocarpus* peak follows the *Hagenia* decline much faster than at Karimu Mire. Its age can be estimated at 24 800 BP (Coetzee 1967). This might indicate that the climatic deterioration recorded at Karimu was of limited duration at the lower altitude lake site, and was not sufficiently severe to affect the expanding dry montane forest at Sacred Lake.

The diagram from Karimu Mire records continued development of dry montane forest until well after 24 200 BP, followed by a relatively rapid decline into cold, dry, glacial conditions when the vegetation belts were again severely depressed on the Aberdare Range. The lowering of the vegetation belt was parallelled at Sacred Lake, where cold, dry conditions are calculated to have reached their maximum between 21 800 and 17 000 BP.

The marine cores taken off the coast of Senegal record a semiarid period in West Africa at about 23 000 BP, which, while drier than today, was warmer and wetter than the very arid period which followed.

Further radiocarbon determinations from the Aberdare site should allow a more accurate definition of conditions in this 'interstadial' period.

LACUSTRINE EVIDENCE

Lake basins with detailed chronologies

The longest and most fully dated Late Quaternary sequence is derived from the Abhé Basin where diatom and sedimentological analyses of a 50 m core taken near the edge of the present lake have been supplemented by detailed studies of a large number of surface exposures (Gasse 1975, 1977, Gasse & Delibrias 1977). The relevant section of the core extends from 21.08 to 11.0 m. It is controlled by a total of six conformable ^{14}C ages, three on organic matter (not charcoal as stated by Butzer (1979) and three on carbonate. A further eight dates have been obtained from the basin at large. Despite problems with the dating of indurated limestone bands in the lake sediments, which appear to have been deposited by circulating geothermal waters (Gasse & Delibrias 1977), the overall sequence yields a highly consistent picture.

A major lacustrine highstand, Abhé III, is represented by clayey, calcareous diatomites in the Abhé core. The preceding maximum, Abhé II, lies beyond the range of ^{14}C, although a questionable date of 39 000 ± 4 000 BP was obtained on organic matter from the Abhé II regressive deposits (Gasse & Delibrias 1977). Two probable minimum ages of 30 600 and 31 000 BP on carbonate also relate to the Abhé II regression. The onset of the Abhé III phase is not precisely fixed. It clearly predates a measured age of 26 900 ± 700 BP on shell from 140 m above present lake level. The end of the Abhé III transgression is dated 25 600 ± 700 BP on organic matter from the core. The timing of the lake-level maximum is supported by six ^{14}C assays on shell and carbonate ranging from 25 100 to 23 190 BP. It seems highly improbable that such a close grouping of dates would be produced by contamination of samples too old to date by ^{14}C. A further three dates, on shell, soil humus and lacustrine carbonate respectively, suggest that a minor recession followed by a temporary recovery occurred around 21 000 BP, that the lake had partially desiccated by 19 000 BP, and that it dried out completely about 17 000 BP.

Environmental conditions during the Abhé III maximum have been reconstructed in detail by Gasse (Gasse 1975, 1977, Gasse & Delibrias 1977). The lake reached its highest level during the entire Late Quaternary: 170 m above the present water surface. Its diatom assemblages were dominated by *Melosira granulata* v. *valida* and other tropical planktonic *Melosira* species which are typical of large dilute lakes of moderate alkalinity. 'Temperate' diatom species were present during the transgressive phase before 25 000 BP and after 21 000 BP, and it is interesting to note that the 'tropical' species which dominated between 25 000 and 21 000 BP are capable of supporting cooler water temperatures than today (Gasse 1980). However, studies of modern lacustrine diatom floras being carried out by Gasse (personal communication) indicate that water balance and water chemistry are much stronger ecological controls than temperature, so that the above conclusions should be regarded as tentative.

A very similar sequence of events has been reconstructed in the Ziway-Shala Basin, Ethiopia (8°N). Surface sections along the Bulbula River expose a great thickness of Upper Quaternary deposits. These are informally termed the Bulbula formation (Street 1979). Up to 22.7 m of diatomaceous lacustrine sands, muds and marls rest on a basal cobble gravel, and are overlain by 13 m of air-fall pumice beds (the Abernosa pumice member) and then by lake beds of latest Pleistocene to mid-Holocene age. The Late Pleistocene lacustral phase represented by the diatomites succeeded a prolonged period of low lake levels. After an initial minor oscillation during which the lake remained brackish (phase Bulbula I, now renamed Ziway-Shala II), there was a short interval of soil formation. This is dated 27 050 ± 1 540 BP on disseminated charcoal (Gasse & Street 1978). The main lacustral phase which followed (Ziway-Shala III) culminated in a maximum at least 83 m above modern base level. Its deposits have been dated 24 000 ± 750 BP at the base and 22 000 ± 650 BP at the top (Street in press). Both ages were derived from low-Mg calcite marls and are therefore likely to be too young, although thin-section studies showed that recrystallization of the original micrite is minimal (Street 1979). A third date of 24 590 ± 550 BP (HAR-2788) has recently been obtained from an ostracod marl which marks the final regression of the Ziway-Shala III lake. In view of this conflict, only the basal charcoal date will provisionally be accepted as reliable.

During the Ziway-Shala III optimum, the diatom flora of the lake was dominated by *Cyclotella ocellata*, *Fragilaria brevistriata* and *Melosira granulata* and/or its varieties, associated with small numbers of 'cold-water' species (Gasse & Descourtieux 1979, Descourtieux-Coqueugnoit 1979). This is essentially the same phytoplanktonic type as in Lake Abhé III. It also records a dilute, moderately alkaline lake. The similarity between the Ziway-Shala III flora and that of phase Ziway-Shala VI (6 500-4 800 BP) suggests that the Late Pleistocene lake reached overflow level.

500 km to the south-west of Lake Ziway-Shala, in the Turkana Basin, where intensive stratigraphic investigations have been carried out by Butzer (Butzer *et al.* 1969, Butzer 1976), no evidence for an upper Late Pleistocene highstand has emerged. The littoral, deltaic and fluvial beds of the Kibish Formation are horizontal and tectonically undisturbed. There is a major disconformity between member III, which yielded an age of >37 000 BP near the top, and the base of member IV, dated 9 500 BP. This stratigraphic gap is interpreted as the result of a period of low lake level; a conclusion which is supported by the local development of carbonate horizons and desert varnish. Such purely negative evidence, however, remains rather unsatisfactory, and should ideally be supported by analyses of cores from the sediments underlying the present lake.

Other well dated sequences have been obtained from Lakes Mobutu (1.5°N) and Manyara (4°S) in equatorial Africa. In both cases, the time framework is based on organic matter from cores, and should therefore be trustworthy. The pattern of fluctuations experienced during the period 30 000-20 000 BP was almost the inverse of Lakes Abhé and Ziway-Shala. Mobutu was more dilute than present between 28 000 and 25 000 BP, but then dropped below its outlet level and entered a period of enhanced salinity lasting until 18 000 BP (Holdship 1976). Lake Manyara was more saline than today throughout the period 45 000-22 000 BP, except for a brief episode of greater dilution which culminated between 27 000 and 26 500 BP. From 22 000 BP onwards its waters again freshened, leading into a major lacustral phase which lasted from 19 400 to 16 000 BP.

In the southern Sahara, the only area where deposits dating from the period 30 000-20 000 BP have been studied in detail is the Chad Basin. Fourteen finite ages greater than 20 000 y, all on carbonates and many with large standard errors, form the basis of the chronology (Faure *et al.* 1963, Servant 1963). During the Ghazalian period there were two lacustral phases, of which the first is believed to have commenced around 38 000 BP and the second around 30 000 BP (Servant-Vildary 1979, Durand & Mathieu 1979). These dates should, of course, be regarded as approximate. The Ghazalian deposits consist of isolated outliers or lenses of calcareous sediments interbedded with aeolian sands. They indicate the presence of numerous small, independent lakes and marshes initiated by a rise in the regional water table in the dunes. These lakes were generally shallow and choked with aquatic vegetation. They experienced frequent temporary spells of desiccation. Extensive areas of open water only developed in the Bahr el Ghazal area (Angela Kete) during the second lacustral phase. The base of the deposits at Angela Kete yielded a measured age of 28 800 ± 1 000 BP (Servant 1973). Sedimentation was continuous between the levels dated 25 600 ± 800 and 20 000-18 000 BP, after which the lake dried up (Servant-Vildary 1979, Servant & Servant-Vildary 1980). Durand & Mathieu (1979) place the lake-level maximum at 22 000 BP.

During the period dated 25 600-22 400 BP, the diatom associations at Angela Kete contain a significant number of temperate-oligotrophic species. The latter decline in number above the 22 400 BP level and disappear before the final desiccation of the lake (Servant & Servant-Vildary 1980). This change led the Servants to postulate cooler climatic conditions before 22 400 BP, due to the frequent advection of polar air masses from the north of the Sahara. Once again, however, there is a possibility that the dominant ecological factor controlling the abundance of the so-called 'cold' taxa was in fact water chemistry.

Other lake basins

Evidence relating to the period 30 000-20 000 BP is not confined to basins with multiple ^{14}C dates, although clearly where only one or two measured ages are available there is much more likelihood of error. In general, we regard unsupported dates > 25 000 BP as unreliable. Many carbonate dates in the range 25 000-20 000 BP may also be too young, although shell is probably somewhat less susceptible than marl in this respect.

In Ethiopia, lacustrine sequences similar to Lake Abhé are also found in the Asal and Hanlé-Dobi grabens (12°N). In the latter, the lake-level maximum was dated 23 600 ± 650 BP and the diatom floras are almost identical to phase Abhé III. Undated Late Pleistocene lake and marsh sediments associated with Middle Stone Age artifacts are found at a number of archaeological sites in the Southern Afar (Gasse et al. 1980).

Diatomaceous sediments strongly resembling those in the Ziway-Shala Basin outcrop on the slopes above Lake Nakuru, Kenya (0°S). Two, or in places three, diatomaceous units are interbedded with colluvial and alluvial deposits containing Middle Stone Age or 'Kenya Stillbay' artifacts (Isaac 1976). The diatomites seem to represent up to three widely spaced lake-level maxima which may have reached overflow level. Carbonate concretions in the uppermost unit yielded a probable minimum age of 21 000 ± 420 BP. This highstand was succeeded by a long period of gully incision and piedmont alluviation. A date of 19 000 ± 400 BP on charcoal from the overlying fluvial depo-

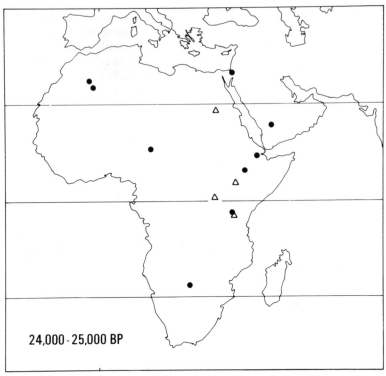

Figure 3. Spatial pattern of lake levels during selected time periods.
Black dots: lake status high; Open circles: lake status intermediate; Triangles: lake
status low; For the definition of lake status, see Street & Grove (1979)

sits implies that the lake was by then very low (Butzer *et al.* 1972).

Other relevant dates on Late Pleistocene lake sediments include the following: two measured ages of 27 900 and 24 400 BP on ostracod-rich deposits
surrounding Ngorongoro Crater, Tanzania (3°S) (Hay 1976); a single date of
23 595 ± 1 800 BP on carbonate encrusting a high shoreline of the Sebkha de
Chemchane, Mauritania (21°N) (Chamard 1973); and a pair of samples of calcareous lake mud within the High Terrace of the Bardai area, Tibesti (21.5°N)
which yielded dates of 24 865 ± 435 and 24 045 ± 410 BP (Jäkel 1979). All
of these measured ages are probably minima. No trace of lake deposits attri-
butable to the period 30 000-20 000 BP has been found in the Egyptian
Sahara although strong deflation in that area militates against the survival of
fine-grained sediments (Wendorf *et al.* 1977).

Spatial pattern

The geographical distribution of lakes which experienced Late Pleistocene

68

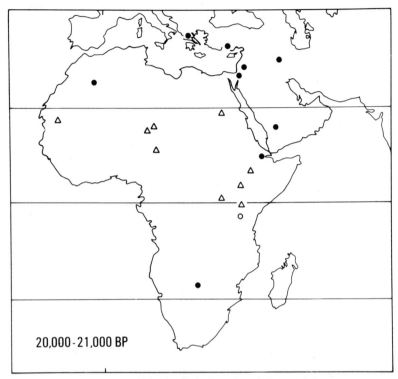

Figure 4. Spatial pattern of lake levels during selected time periods.
Black dots: lake status high; Open circles: lake status intermediate; Triangles: lake
status low; For the definition of lake status, see Street & Grove (1979)

highstands is illustrated in figures 3-5. We have not included any maps for the
period prior to 25 000 BP, when the radiocarbon framework becomes un-
acceptably imprecise and unreliable. For the same reason, figure 3 should be
regarded as an indication of the spread of published data rather than as a
firmly substantiated picture.

The conclusions which can be drawn in the light of figures 3-5 are as follows: there is an increasing amount of stratigraphic evidence for upper Late
Pleistocene highstands in tropical Africa. If the published radiocarbon chronologies are broadly accepted, but putting most weight on sites with multiple
^{14}C dates, it appears that there was an extensive lacustral phase in northern
intertropical Africa which culminated between about 25 000 and 22 000 BP.
The belt of high levels stretched from just south of the equator to at least
22°N. It also extended eastwards into Saudi Arabia (figures 3-4). Sites with
good stratigraphic resolution, such as Abhé, Ziway-Shala and Nakuru, suggest
that there were in fact two or three major maxima during the period of inte-

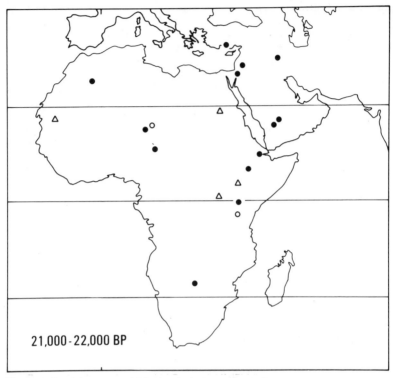

Figure 5. Spatial pattern of lake levels during selected time periods.
Black dots: lake status high; Open circles: lake status intermediate; Triangles: lake
status low; For the definition of lake status, see Street & Grove (1979)

rest, but it may never be possible to resolve these fully using ^{14}C analyses. In
contrast, the pattern of fluctuations registered by some of the equatorial
lakes appears to have been significantly different. And whereas Lake Makga-
dikgadi (20°S) shows strong similarities to the Abhé record (Grove 1978,
Heine 1978), there is insufficient evidence to reach any conclusions about
the pattern of fluctuations in the southern tropics.

By 21 000 BP, a strong drying trend had set in, resulting in the rapid
shrinkage or even complete desiccation of the remaining lakes (figure 5).

OCEANIC EVIDENCE FOR FLUCTUATIONS IN RIVER DISCHARGE

One of Professor Butzer's major criticisms of earlier climatic reconstructions
based on lake-level evidence is that no account has been taken of the record
of fluvial erosion and aggradation (Butzer personal communication). In part,

this can be attributed to the lack of consensus among fluvial geomorphologists, particularly with regard to the Nile (Williams & Faure 1980). The controversy surrounding the interpretation of fluvial sequences has centered on their interpretation in terms of fluctuations in discharge and/or sediment transport. Many existing studies have tended to ignore the crucial effects of varying sediment supply on fluvial dynamics and channel morphology (Schumm 1977), concentrating instead on the role of discharge. The interaction between the outputs of different tributary basins has also not been fully considered (Adamson *et al.* 1980).

This confused situation is fortunately improving. High-resolution marine cores taken close to the mouths of the major river systems now permit an independent evaluation of changes in freshwater discharge into the oceans. Stanley & Maldonado (1977, 1979) have studied an extensive suite of cores from the Nile Cone. These display a cyclical alternation of muds and calcareous oozes with thin organic bands known as sapropels. The latter are thought to represent periods of high freshwater input into the eastern Mediterranean, resulting in retarded vertical mixing and oxygen depletion in the bottom water (Luz 1979). The youngest sapropel, S_1, is dated $>8\,200$-$6\,000$ BP on organic matter (Stanley & Maldonado 1977). Although alternative sources for the freshwater input include the Black Sea and the northern Mediterranean rivers, S_1 is increasingly being attributed by oceanographers to a Holocene maximum in discharge from the Nile (Vergnaud-Grazzini personal communication). The second sapropel S_2 is less well developed and has a more restricted distribution within the eastern Mediterranean (Stanley 1978, Williams *et al.* 1978, Cita *et al.* 1977). It is dated ca.25 000-23 000 BP (Stanley & Maldonado 1977). Sapropel S_2 provides strong circumstantial evidence for a pulse of increased Nile runoff into the eastern Mediterranean between 25 000 and 23 000 BP.

Another recent source of information has been core KW31 from the outer Niger delta (Pastouret *et al.* 1978). A record of freshwater discharge from the Niger is provided by the $\delta^{18}O$ curve derived from planktonic foraminifera (*G.ruber*). This differs systematically from the $\delta^{18}O$ curve for benthic foraminifera. The divergences are thought to represent periods of strong dilution of the surface waters by terrestrial runoff. Major discharge maxima are believed to have occurred from 4 500 to 11 500 BP, from 11 800 to 13 000 BP and around 22 000 ± 2 000 BP. The close agreement between the first two and the lake-level curve for Chad (Servant & Servant-Vildary 1980) suggests that the 22 000 BP peak also represents a lacustral episode. If so, then its magnitude was similar to the oscillation centered on 12 000 BP (Pastouret *et al.* 1978, figure 2) rather than to conditions during the early and mid-Holocene.

High-resolution deep sea cores also permit us to assess the wider climatic setting of the 25 000-22 000 BP wet phase. This is situated right on the boundary between isotope stages 2 and 3, now dated at approximately 24 000-23 000 BP (Pastouret *et al.* 1978, Lutze *et al.* 1979). Any search for the underlying climatic mechanisms must therefore take into account that this wet episode occurred during a phase of rapid transition from interstadial to full-glacial conditions. There is little indication that any localized warming of ocean temperatures occurred in the waters around Africa during the period of interest (Thiede 1977, Lutze *et al.* 1979, Hutson 1980).

CONCLUSIONS

In this paper, we have presented evidence for a forest period in East Africa which culminated after 24 000 BP. This is supported by indications of higher levels in many tropical closed-basin lakes north of the equator, bearing in mind the less satisfactory nature of their dating frameworks. The lake-level data suggest a maximum of wetness between 25 000 and 22 000 BP. Further confirmation has also come from recent studies of deep sea cores off the mouths of the Niger and the Nile.

How far can we interpret this moist phase in climatic terms? The pollen data indicate that dry forest similar to today was present on the Aberdares, implying that conditions were less humid than during the early to mid-Holocene forest period. The lake-level evidence is hard to interpret, because of the well-known ambiguity of water-budget calculations for periods cooler than present (Galloway 1970, Coventry 1976, Brakenridge 1978). In purely relative terms, the upper Late Pleistocene maximum was less pronounced than the early and mid-Holocene maxima in many basins, for example Chad, where the extent of lakes seems to have been more comparable with the period 12 000-11 000 BP (Servant & Servant-Vildary 1980). Servant-Vildary (1979) has therefore concluded that the Ghazalian climate was relatively dry, although more humid than at present. The greatest impact of the upper Late Pleistocene wet phase was felt in the Ethiopian Highlands. Lakes Abhé and Ziway-Shala became more dilute than at any other time during the Late Quaternary (Gasse 1980). Lake Abhé also reached its greatest surface extent (Gasse & Street 1978).

The question therefore remains as to how far the lake-level maximum can be attributed to lower temperatures and how far to increased rainfall or to other factors. The circumstantial diatom evidence for cooler conditions in Chad and Ethiopia has already been mentioned. This is supported by the moderate lowering of the vegetation zones suggested in the Aberdares. Unfortunately, the contemporary extent of glaciation remains entirely unknown. If decreased evaporation is to be held entirely responsible for the expanded lakes north of the equator, however, this implies that temperatures were lowered by a greater amount in the region of Lakes Mobutu and Manyara than further north.

Another factor which should be taken into account is the effect of vegetation cover on runoff. At present, the Kenyan and Ethiopian Rift lakes are fed mainly by rivers draining highland areas above 2 000 m. If, as the Aberdare data suggest, the present montane forest belt was occupied by relatively open

72

woodland during the early part of the period of interest, then the runoff coefficient (the proportion of rainfall becoming runoff) was probably somewhat higher than today. Studies of lacustrine sediments in Chad and Ethiopia suggest that the contribution of river runoff to the water and sediment budgets of the lakes during the interval 30 000-20 000 BP was highly significant (Servant & Servant-Vildary 1980, Gasse & Street 1978).

We therefore tentatively conclude that the upper Late Pleistocene wet phase in northern intertropical Africa resulted from a combination of cooler temperatures and slightly greater precipitation than today. Its impact may have been enhanced by higher runoff rates and possibly by increased cloudiness. Environmental conditions in many areas most closely resembled the period 12 000-11 000 BP, which was again a time of climatic transition. The spatial distribution of high lake levels also invites comparison with the period 9 000-8 000 BP (Street & Grove 1979), although the impact of the change in water balance was less dramatic. More long sequences with ^{14}C dates on organic carbon will, however, be needed before further progress can be made in understanding the spatial and temporal pattern of events between 30 000-20 000 BP.

ACKNOWLEDGEMENTS

We thank G.Delibrias, R.Gillespie, D.D.Harkness and R.L.Otlet for carrying out the radiocarbon determinations, and A.C.Hamilton and F.Gasse for many stimulating discussions. A.C.Hamilton also helped to collect some of the cores from Karimu Mire. We thank the Natural Environment Research Council, the Royal Society and the Department of Geography, Cambridge University, for financial support.

REFERENCES

Adamson, D.A., F.Gasse, F.A.Street & M.A.J.Williams 1980. Late Quaternary history of the Nile. *Nature* 288: 50-55.
Brakenridge, G.R. 1978. Evidence for a cold, dry full-glacial in the American Southwest. *Quat. Res.* 9: 22-40.
Butzer, K.W. 1979. Climatic patterns in an unglaciated continent. *Geogr. Mag.* 51: 201-208.
Butzer, K.W. 1976, The Mursi, Nkalabong and Kibish Formations, Lower Omo Basin, Ethiopia. In: Y.Coppens, F.C.Howell, G.Ll.Isaac & R.E.F.Leakey (eds.), *Earliest Man and Environments in the Lake Rudolf Basin.* Chicago Press, Chicago: 12-23.
Butzer, K.W., F.H.Brown & D.L.Thurber 1969. Horizontal sediments of the Lower Omo Valley: the Kibish Formation. *Quaternaria* 11: 15-30.
Butzer, K.W., G.Ll.Isaac, J.L.Richardson & C.Washbourn-Kamau 1972. Radiocarbon dating of East African lake levels. *Science* 175: 1069-1076.
Chamard, P. 1973. Monographie d'une sebkha continentale du Sud-Ouest saharien: la sebkha de Chemchane (Adrar de Mauritanie). *Bull. Inst. fond. Afrique noire, Sénégal* 35: 207-243.
Cita, M.B., C.Vergnaud-Grazzini, C.Robert, H.Chamley, N.Ciaranfi & S.d'Onofrio 1977. Palaeoclimatic record of a long deep sea core from the Eastern Mediterranean. *Quat. Res.* 8: 205-235.

Coetzee, J.A. 1967. Pollen analytical studies in East and Southern Africa. *Palaeoecology of Africa* 3: 146pp.

Coventry, R.J. 1976. Abandoned shorelines and the Late Quaternary history of Lake George, New South Wales. *J. Geol. Soc. Australia* 23: 249-273.

Descourtieux-Coqueugniot, C. 1979. *Les diatomées du sondage du lac Abiyata (Ethiopie): systématique et paléoécologie.* Thèse de 3ᵉ cycle, Université Paris-VI, 76pp.

Durand, A. & P.Mathieu 1979. Essai de reconstitution de l'évolution paléoclimatique du bassin tchadien au Pléistocène supérieur à partir de l'étude des formations fluvio-deltaïques du fleuve Chari. *ASEQUA Bull. de Liaison* 56/57: 69-71.

Faure, H. 1969. Lacs quaternaires du Sahara. *Mitt. Int. Verein. Limnol.* 17: 131-146.

Faure, H., E.Manguin & R.Nydal 1963. Formations lacustres du Quaternaire supérieur du Niger oriental: diatomites et âges absolus. *Bull. Bur. Rech. Géol. Min., Paris* 3: 41-63.

Flenley, J.R. 1973. The use of modern pollen rain samples in the study of the vegetational history of tropical regions. In: H.J.B.Birks & R.G.West (eds.), *Quaternary Plant Ecology,* Blackwells,Oxford: 131-141.

Flenley, J.R. 1979. *The equatorial rain forest: a geological history.* Butterworths, London, 162pp.

Galloway, R.W. 1970. The full-glacial climate in the south-western United States. *Ann. Assoc. Am. Geogr.* 60: 245-256.

Gasse, F. 1975. *L'évolution des lacs de l'Afar Central (Ethiopie et TFAI) du Plio-Pléistocène à l'Actuel: Reconstitution des paléomilieux lacustres à partir de l'étude des diatomées.* Thèse d'Etat, Université Paris-VI, 3 vols.

Gasse, F. 1977. Evolution of Lake Abhé (Ethiopia and TFAI) from 70 000 BP. *Nature* 265: 42-45.

Gasse, F. 1980. Late Quaternary changes in lake levels and diatom assemblages on the southeastern margin of the Sahara. *Palaeoecology of Africa* 12: 333-350.

Gasse, F. & G.Delibrias 1977. Les lacs de l'Afar Central (Ethiopie et TFAI) au Pléistocène supérieur. In: S.Horie (ed.), *Paleolimnology of Lake Biwa and the Japanese Pleistocene.* Kyoto, 4: 529-575.

Gasse, F. & C.Descourtieux 1979. Diatomées et évolution de trois milieux éthiopiens d'altitude différente, au cours du Quaternaire supérieur. *Palaeoecology of Africa* 11: 117-134.

Gasse, F., P.Rognon & F.A.Street 1980. Quaternary history of the Afar and Ethiopian Rift Lakes. In: M.A.J.Williams & H.Faure (eds.), *The Sahara and the Nile* Balkema, Rotterdam: 361-400.

Gasse, F. & F.A.Street 1978. Late Quaternary lake level fluctuations and environments of the northern Rift Valley and Afar region (Ethiopia and Djibouti). *Palaeogeogr., Palaeoclimatol., Palaeoecol.* 24: 279-325.

Grove, A.T. 1978. Late Quaternary climatic change and the conditions for current erosion in Africa. *Géo-Eco-Trop.* 2: 291-300.

Hamilton, A.C. 1972. The interpretation of pollen diagrams from highland Uganda. *Palaeoecology of Africa* 7: 45-149.

Hamilton, A.C. & R.A.Perrott 1978. Date of deglacierization of Mount Elgon. *Nature* 273: 49.

Hamilton, A.C. & R.A.Perrott 1979. Aspects of the glaciation of Mount Elgon, East Africa. *Palaeoecology of Africa* 11: 153-161.

Hamilton, A.C. & R.A.Perrott in press. Modern pollen deposition on Mt Elgon, Kenya. *Pollen et Spores.*

Hay, R.L. 1976. *Geology of the Olduvai Gorge – a study of sedimentation in a semiarid basin.* University of California Press, Berkeley, 205pp.

Heine, K. 1978. Radiocarbon chronology of Late Quaternary lakes in the Kalahari, southern Africa. *Catena* 5: 145-149.

Holdship, S.A. 1976. *The paleolimnology of Lake Manyara, Tanzania: a diatom analysis of a 56 meter sediment core.* PhD thesis, Duke University, 121pp.

Hutson, W.H. 1980. The Agulhas Current during the Late Pleistocene: analysis of modern faunal analogs. *Science* 207: 64-66.

Isaac, G.Ll. 1976. A preliminary report on stratigraphic studies in the Nakuru Basin, Kenya. In: B.Abebe, J.Chavaillon & J.E.G.Sutton (eds.), *Proceedings of the VI Panafrican Congress of Prehistory and Quaternary Studies*. Ethiopian Ministry of Culture, Addis Ababa: 409-411.

Jäkel, D. 1979. Runoff and fluvial formation processes in the Tibesti Mountains as indicators of climatic history in the Central Sahara during the Late Pleistocene and Holocene. *Palaeoecology of Africa* 11: 13-39.

Lutze, G.F., M.Sarnthein, B.Koopmann, U.Pflaumann, H.Erlenkeuser & J.Thiede 1979. 36. Meteor cores 12309: Late Pleistocene reference section for interpretation of the Neogene of site 397. In: U.von Rad, W.B.F.Ryan *et al.* (eds.), *Initial Reports of the Deep Sea Drilling Project* 47.

Luz, B. 1979. Palaeoceanography of the post-glacial Eastern Mediterranean. *Nature* 278: 847-848.

Pastouret, L., H.Chamley, G.Delibrias, J.-C.Duplessy & J.Thiede 1978. Late Quaternary climatic changes in western tropical Africa deduced from deep-sea sedimentation off the Niger Delta. *Oceanologica Acta* 1: 217-232.

Rossignol-Strick, M. & D.Duzer 1979a. A Late Quaternary continuous climatic record from palynology of three marine cores off Senegal. *Palaeoecology of Africa* 11: 185-188.

Rossignol-Strick, M. & D.Duzer 1979b. West African vegetation and climate since 22 500 BP from deep-sea cores palynology. *Pollen et Spores* 21: 105-134.

Schumm, S.A. 1977. *The fluvial system.* John Wiley, New York, 338pp.

Servant, M. 1973. *Séquences continentales et variations climatiques: Evolution du bassin du Tchad au Cénozoïque supérieur.* Thèse d'Etat, ORSTOM, Paris, 348pp.

Servant, M. & S.Servant-Vildary 1980. L'environnement quaternaire du bassin du Tchad. In: M.A.J.Williams & H.Faure (eds.), *The Sahara and the Nile.* Balkema, Rotterdam: 133-162.

Servant-Vildary, S. 1979. Paléolimnologie des lacs du bassin tchadien au Quaternaire récent. *Palaeoecology of Africa* 11: 65-78.

Stanley, D.J. 1978. Ionian Sea sapropel distribution and Late Quaternary palaeoceanographic considerations in the Eastern Mediterranean. *Nature* 274: 149-152.

Stanley, D.J. & A.Maldonado 1977. Nile Cone: Late Quaternary stratigraphy and sediment dispersal. *Nature* 266: 129-135.

Stanley, D.J. & A.Maldonado 1979. Levantine Sea – Nile Cone lithostratigraphic evolution: Quantitative analysis and correlation with paleoclimatic and eustatic oscillations in the Late Quaternary. *Sediment. Geol.* 23: 37-65.

Street, F.A. 1979. *Late Quaternary lakes in the Ziway-Shala Basin, southern Ethiopia.* PhD thesis, Cambridge University.

Street, F.A. in press. Chronology of late Pleistocene and Holocene lake-level fluctuations, Ziway-Shala Basin, Ethiopia. In: *Proceedings of the VIII Panafrican Congress of Prehistory and Quaternary Studies,* Nairobi.

Street, F.A. & A.T.Grove 1979. Global maps of lake-level fluctuations since 30 000 BP. *Quat. Res.* 12: 83-118.

Thiede, J. 1977. Aspects of the variability of the Glacial and Interglacial North Atlantic boundary current (last 150 000 years). *'Meteor' Forsch.-Ergebnisse* C28: 1-36.

Wendorf, F. *et al.* 1977. Late Pleistocene and Recent climatic changes in the Egyptian Sahara. *Geogr. J.* 143: 211-234.

Williams, D.F., R.C.Thunell & J.P.Kennett 1978. Periodic freshwater flooding and stagnation of the eastern Mediterranean during the Late Quaternary. *Science* 201: 252-254.

Williams, M.A.J. & H.Faure (eds.) 1980. *The Sahara and the Nile.* Balkema, Rotterdam, 607pp.

A HIGH ALTITUDE POLLEN DIAGRAM FROM MOUNT KENYA: ITS IMPLICATIONS FOR THE HISTORY OF GLACIATION

R. A. PERROTT
School of Biological & Environmental Studies, University of Ulster,
Coloraine, Northern Ireland

In 1978, during a reconnaissance survey of the glaciated and formerly glaciated regions of Mount Kenya, a number of cores were collected for subsequent analysis. The pollen diagram presented here was derived from sediments collected from behind a previously unrecorded moraine at an altitude of 4 265 m in the Hobley Valley. This diagram enables us to determine a minimum age for glacial retreat from moraines at this altitude.

A pit was excavated to bedrock (52.5 cm) in peat deposits immediately up-valley from a moraine (fig.1), which correlates in altitude and appearance with the moraines termed 'Stage IV' in the Mackinder, Hinde, Gorges, Teleki and Höhnel Valleys (Baker 1967).

In the field the sediment stratigraphy was noted as follows:

0 - 14 cm	brown peat with numerous fibrous rootlets.
14 - 28.5 cm	red-brown peat, numerous gritty particles, few macrofossils.
28.5 - 33.5 cm	grey band, light grey at top gradually becoming darker to almost black at 33 cm, smooth with very fine gritty particles, contains angular mineral particles ≤4 cm.
33.5 - 39 cm	brown peat, numerous sedge-like macrofossils, some angular mineral particles.
39 - 52.5 cm	grey-grown peat, numerous sedge-like macrofossils, angular mineral particles ≤1.5 cm becoming very frequent towards base.

Samples for analysis were collected at 1 cm increments throughout the profile.

The mire lies within the 'Afroalpine' zone (Hedberg 1951, 1964, Coe 1967) and is surrounded by a plant community dominated by tussock grasses and species of the genera *Senecio, Lobelia, Alchemilla* and *Helichrysum.* Cyperaceae are present on the mire surface and wetter areas of the valley floor. The Hobley Valley is not served by any of the tracks giving access to the peak area and its inaccessibility is likely to have ensured that there will have been little, if any, anthropogenic impact on the vegetation. The modern upper limit of the montane forest lies ca. 1 000 m in altitude below the mire.

The pollen diagram (fig.2) has been constructed with the prime aim of establishing a minimum date of glacial retreat from the moraine. With this in mind the diagram has been divided into two zones labelled A and B on the basis of the marked increase in *Podocarpus* pollen at 36 cm depth in the pro-

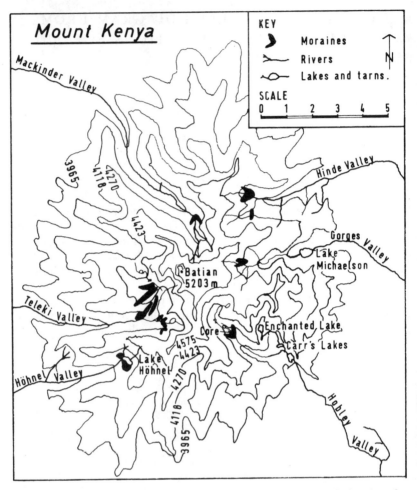

Figure 1. Simplified map of the summit area of Mount Kenya showing the location of the moraine and core site at 4 265 m altitude in the Hobley Valley and the moraines termed 'Stage IV' (Baker 1967) in the Mackinder, Hinde, Gorges, Höhnel and Teleki Valleys.

file. This increase has been observed in pollen diagrams from sites on a number of East African mountains, most notably, Sacred Lake, Mount Kenya (Coetzee 1967), Lake Mahoma, Ruwenzori Range (Livingstone 1967), Lake Kimilili, Mount Elgon (Hamilton & Perrott unpublished) and Mount Satima, Aberdare Range (Perrott unpublished). It is normally associated with increases in the pollen representation by other trees of the drier montane forest type, such as, *Juniperus procera* and *Olea* sp., which I believe marks the onset of drier con-

Table 1. Extrapolated dates for the *Podocarpus* rise in pollen diagrams from montane sites in East Africa.

Location	Depth	^{14}C date	Extrapolated date *Podocarpus* rise
Lake Kimilili,	140 - 155 cm	3 080 ± 180 BP	
Mount Elgon.	190 cm	————————	3 720 BP
4 150 m	215 - 225 cm	4 170 ± 280 BP	
Sacred Lake,	670 - 680 cm	3 285 ± 60 BP	
Mount Kenya.	690 cm	————————	3 805 BP
2 400 m	880 - 890 cm	10 560 ± 65 BP	
Lake Mahoma,	Surface		
Ruwenzori Mts.	175 cm	————————	3 553 BP
2 960 m	225 - 250 cm	4 670 ± 80 BP	

ditions in East Africa. This increase in *Podocarpus* pollen can be fairly reliably dated at a number of locations as illustrated in Table 1.

The most precise dating of the *Podocarpus* rise is at ca. 3 720 BP, at Kimilili Lake, Mount Elgon which is closest in altitude and aspect to the Hobley Valley mire. This date (3 270 BP) which is very close to the mean date (3 693 BP) for the *Podocarpus* rise at the sites mentioned in table 1, will be used as the basis for the calculation of a minimum date for glacial retreat from the Hobley Valley moraine.

If 3 270 BP is taken to be the date for the *Podocarpus* rise at 36 cm depth in the Hobley Valley mire, then the rate of sediment accumulation between the present day and 3 720 BP will have been 1 cm in 103.3 years. If sedimentation rates are assumed to have been constant to the base of the core, then

Table 2. Sedimentation rates for selected sites in East Africa.

Location	Period	Deposition rate
Muchoya Swamp,	0 - 6 570 BP	1 cm in 13.8 years
Uganda.	6 570 - 9 340 BP	1 cm in 18.5 years
(Morrison 1968)	9 340 - 12 890 BP	1 cm in 23.7 years
Mahoma Lake,	0 - 4 670 BP	1 cm in 20.3 years
Ruwenzori Mountains.	4 670 - 12 700 BP	1 cm in 36.5 years
(Livingstone 1967)		
Sacred Lake,	0 - 3 285 BP	1 cm in 14.6 years
Mount Kenya.	3 285 - 10 560 BP	1 cm in 38.6 years
(Coetzee 1967)		
Kimilili Lake,	0 - 4 170 BP	1 cm in 18.9 years
Mount Elgon.	4 170 - 6 860 BP	1 cm in 29.1 years
(Hamilton & Perrott, unpublished)		

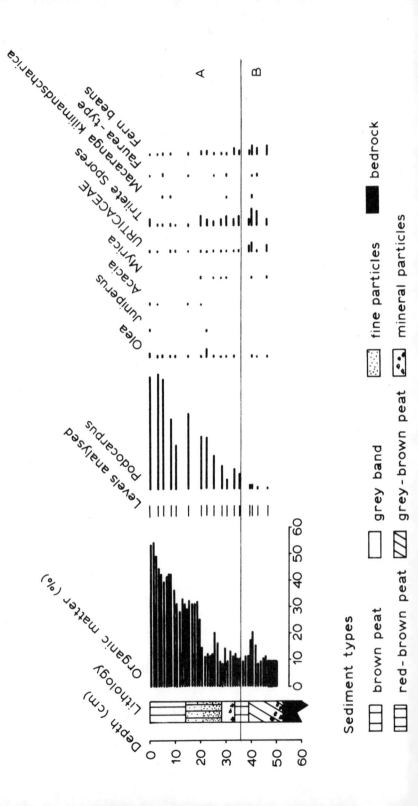

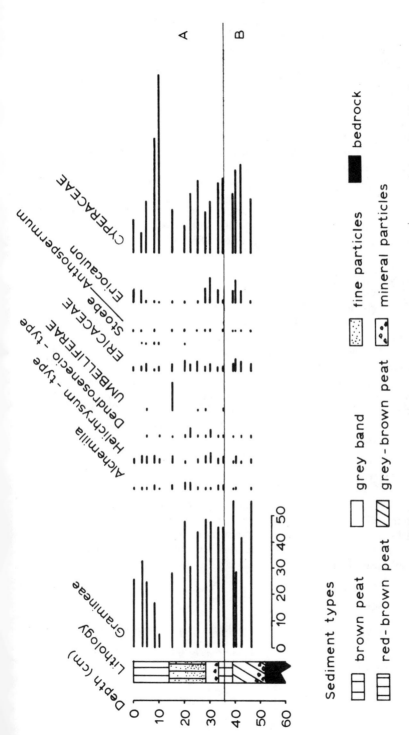

Figure 2. A pollen diagram from the Hobley Valley, Mount Kenya. Pollen sum = total pollen. All counts ≥ 500 grains.

the minimum date for glacial retreat from the Hobley Valley moraine can be calculated to be prior to ca. 5 425 BP.

However, the evidence for rates of sediment accumulation at other montane sites in East Africa indicates that the rate was slower prior to the onset of drier conditions; just how much slower depends on the site chosen as illustrated in table 2.

The rate of sediment accumulation varies considerably from site to site, but again Lake Kimilili with a sedimentation rate after 4170 BP of approximately 1.5 times that before that date, provides the best analog. By assuming that the sediment accumulation rate after 3720 BP was 1.5 times that before that date we get a minimum date for glacial retreat from the Hobley Valley moraine in the range ca. 5425 BP (constant sedimentation rate) to ca. 6277 BP.

It is interesting to note that Coetzee (1967) found evidence in her pollen diagram from Sacred Lake for a colder, moister period of ± 1 000 years duration centered on a calculated age of ca. 5 860 BP.

CONCLUSIONS

(1) The glacier that constructed a moraine at an altitude of 4 265 m in the Hobley Valley retreated prior to ca. 5 425/6 277 BP depending on whether a constant or variable sedimentation rate is used to calculate the age.

(2) The glacial episode that was responsible for the Hobley Valley moraine probably also constructed the moraines in the Mackinder, Hinde, Gorges, Höhnel and Teleki Valleys termed 'constructional moraines of Stage IV' (Baker 1967). If this hypothesis is correct then Baker's intuitive correlation between his Stage IV moraines and the Daun stage (ca. 12 000 BP) in the Alps is incorrect, with obvious consequences for the accepted chronology of the Mount Kenya glacial sequence.

ACKNOWLEDGEMENTS

This work was supported by a grant from the Natural Environment Research Council (UK). I am grateful to the Government of Kenya and the Kenya Parks Department for permission to work on Mount Kenya and to the Botany Department, University of Nairobi for their support. Dr F.A.Street-Perrott kindly read the manuscript.

REFERENCES

Baker, B.H. 1967. Geology of the Mount Kenya area. *Kenya Geological Survey,* Report No.79, 78pp.
Coe, M.J. 1967. *The ecology of the Alpine Zone of Mount Kenya.* W.Junk, The Hague. 136pp.

Coetzee, J.A. 1967. Pollen analytical studies in East and southern Africa. *Palaeoecology of Africa* 3:1-146.

Hamilton, A.C. & R.A.Perrott. Unpublished pollen diagrams.

Hedberg, O. 1951. Vegetation belts of the East African mountains. *Svensk bot. Tidskr.* 45:140-202.

Hedberg, O. 1964. Features of Afroalpine Plant Ecology. *Acta Phytogeographica Suecica* 49:1-144.

Livingstone, D.A. 1967. Postglacial vegetation of the Ruwenzori Mountains in equatorial Africa. *Ecological Monographs* 37:25-52.

Perrott, R.A. Unpublished pollen diagrams.

RECHERCHES PALYNOLOGIQUES DANS LA REGION DU LAC TURKANA (KENYA)

A. VINCENS
Laboratoire de Géologie du Quaternaire, CNRS, Luminy, Marseille, France

SUMMARY

For the pollen analytical study of plio-pleistocene history of the Lake Turkana region 30 recent surface soil samples and 34 samples from the surface of the lake deposits have been analysed. It appears that the pollen of the montane forest is not dispersed far by wind, while river transport is more effective.

Using this data the following climatic succession can be inferred from 17 fossil pollen spectra:
- from 2.7-2.1 MY: climate warm and humid.
- from 2.1-1.8 MY: climate dry with a tendency towards cooling.
- from 1.8-1.5 MY: climate more humid and still cooler.
Holocene pollen spectra cannot yet be explained in terms of climate.

INTRODUCTION

Les recherches palynologiques entreprises depuis 1976 à l'Est du Lac Turkana font partie d'un programme international: 'Koobi Fora Research Project', dirigé par R.E.F.Leakey & G.Isaac centré sur la reconstitution du paléoenvironnement des Hominidés fossiles découverts depuis plusieurs années dans les dépôts fluvio-lacustres plio-quaternaires de cette région (Leakey & Leakey. 1978). Associée aux autres disciplines, comme la géologie, la sédimentologie et la paléontologie, la palynologie apporte un complément d'informations précieuses pour de telles reconstitutions, en particulier en ce qui concerne le paysage végétal et la climatologie.

Peu de travaux palynologiques ont été effectués en Afrique de l'Est, particulièrement dans les zones sub-désertiques de basse altitude. Les seules données actuellement connues concernent les dépôts pléistocènes du lac Abhé (Ethiopie) (Bonnefille *et al.* sous presse) et les dépôts plio-pléistocènes de la basse vallée de l'Omo (Ethiopie) (Bonnefille 1972, 1976) dont ce travail est le complément.

1. METHODOLOGIE

L'étude palynologique des sédiments plio-quaternaires d'Est Turkana et leur interprétation sont abordées par comparaison avec des données tout à fait nouvelles pour l'Afrique tropicale. L'analyse d'échantillons de surface de sol avait déjà été effectuée par divers auteurs (Bonnefille 1972, 1979; Hamilton 1972). Mais pour la première fois est entreprise celle de nombreux échantillons lacustres actuels répartis sur toute la surface d'un lac.

Les études sur la sédimentation des pollens existent en zones tempérées de l'Hémisphère Nord, en particulier aux Etats-Unis (Davis 1967, 1968; Davis *et al.* 1971; Davis *et al.* 1969) et en Angleterre (Bonny 1978, 1980). Quelques données ont été obtenues dans des régions sub-désertiques, en Israël (Rossignol 1969, Horowitz 1969) et au Tchad (Maley 1972, 1980). En Afrique, il s'agit de la première étude de ce genre.

Le travail sur la pluie et la sédimentation pollinique actuelle dans le bassin du lac Turkana apparait donc doublement important: d'une part par sa nouveauté, d'autre part par le modèle qu'il représente pour l'interprétation des spectres fossiles d'Est Turkana, extraits pour la plupart de dépôts lacustres.

2. PALYNOLOGIE ACTUELLE DANS LE BASSIN DU LAC TURKANA

L'étude de la pluie et sédimentation pollinique actuelles dans le bassin du lac Turkana a donc été abordée sous deux aspects:

2.1. *Analyse d'échantillons de surface*

Au total, 30 échantillons de sol ont été analysés. Ceux-ci avaient été prélevés au cours de 3 missions (1975-1977-1979) aux milieux des principales associations végétales actuelles existant, soit en bordure est du lac (région de Koobi Fora: steppe sub-désertique), soit dans la région (district de Marsabit et de Samburu: savane boisée et forêts d'altitude). Ce travail a permis d'obtenir une représentation pollinique pour chaque association, mais aussi de définir, pour chaque spectre, les taxa les plus représentatifs d'une formation végétale donnée.

Mais le but principal de ce travail était d'évaluer qualitativement et quantitativement le transport à courte, moyenne et longue distance, par voie aérienne de certains taxa, en particulier des éléments de forêts d'altitude, allochtones à la bordure est du Lac Turkana (Bonnefille & Vincens 1977). Les principaux résultats obtenus (tableau 1) montrent que peu de pollens d'éléments de forêts sont transportés par voie aérienne, à moyenne et longue distance, autant en quantité qu'en nombre de taxa différenciés. Au delà de 30 km, les pourcentages enregistrés ne sont jamais supérieurs à 4 %, et seul *Podocarpus* est régulièrement présent dans les spectres.

Tableau 1. Distribution des taxa de forêt d'altitude dans les échantillons de surface de la région du Lac Turkana (% calculés par rapport à Σ P).

Echantillons	Nbre	Localisation/ forêt	%	Nbre de taxa	% des principaux composants rencontrés	
Maralal	1	sous la forêt	54.1	13	Podocarpus	16,4
					Juniperus	11,2
					Olea	10,2
Mt Nyiru	2	lisière de la forêt	13,1	11	Podocarpus	5,8
					Juniperus	0,9
					Olea	2,9
Baragoi	1	30 km	3,9	9	Podocarpus	2,0
					Juniperus	0,5
					Olea	0,3
Transect Marsabit-Koobi Fora	9	30-100 km	< 4	< 6	Podocarpus	< 2,6
					Juniperus	rares
					Olea	
Koobi Fora	11	150 km	< 2	< 3	Podocarpus	< 1,3
					Juniperus	rares
					Olea	

2.2. Analyse de sédiments actuels du Lac Turkana

34 échantillons lacustres actuels ont été analysés. Ceux-ci ont été prélevés soit par sondage ou draguage par R.F.Yuretich (1973-74) et C.Barton (1979). Répartis sur toute la surface du lac, l'étude de tels échantillons nous à permis d'avoir des données précises de la sédimentation pollinique actuelle dans le bassin lacustre de Turkana (Vincens 1979).

Pour chaque taxa et groupe de taxa représentant les différentes associations végétales locales ou régionales entourant le bassin, nous avons pu dresser des cartes de distribution des pourcentages de pollens à la surface du lac. Celles-ci montrent que la végétation locale (steppe sub-désertique) est la mieux traduite par les sédiments de la bordure du lac. Par contre, la végétation régionale (forêts riveraines et de montagne + Typha + Pteridophytes) est la mieux représentée dans les spectres des zones deltaïques des rivières Omo au Nord et Kerio-Turkwel au Sud-Ouest.

De plus, pour une même distance de transport, les rivières alimentant le lac Turkana sont de bien meilleurs vecteurs de transport de pollens que les vents, en particulier en ce qui concerne les taxa de forêt d'altitude, puisque ceux-ci représentent en moyenne 5,6 % du total des pollens comptés dans les sédiments lacustres francs et 13,5 % dans les dépôts deltaïques.

Pour comprendre et interpréter les résultats palynologiques plio-quaternaires d'Est Turkana, il faudra donc tenir compte des divers facteurs qui peuvent affecter la distribution des pollens, en particulier l'environnement sédimentaire des dépôts fossiles analysés et leur position précise dans la paléogéographie.

3. PALYNOLOGIE PLIO-QUATERNAIRE D'EST TURKANA

Ce travail, centré au début sur l'analyse des dépôts d'âge plio-pléistocène dans lesquels se trouvent les plus importants gisements à Hominidés et sites archéologiques d'Est Turkana à été récemment étendu aussi aux dépôts lacustres holocènes.

3.1. *Dépôts plio-pléistocènes: Formation de Koobi Fora*

Les données palynologiques extraites de l'analyse des sédiments affleurant à l'Est du lac Turkana, sont actuellement les plus importantes obtenues en Afrique de l'Est pour le Plio-Pléistocène. Pour une période comprise entre 2.5 et 1.5 MA, 17 spectres ont été obtenus. Une partie des résultats ont déjà fait l'objet de plusieurs publications (Bonnefille 1976, Vincens 1979, Bonnefille & Vincens sous presse) et une première tentative d'interprétation des résultats à pu être faite (Vincens sous presse). Grâce aux variations qualitatives et quantitatives enregistrées dans les spectres fossiles, en particulier par les éléments de forêt d'altitude, nous avons pu mettre en évidence, entre 2.5 et 1.5 MA, trois grands épisodes climatiques distincts:

3.1.1. *Période antérieure à 2.1 MA.* Elle est représentée par 4 spectres caractérisés par la présence de taxa de forêt d'altitude particuliers tels que *Ilex, Ericaceae, Myricaceae, Pteridophytes* et pourcentages supérieurs à ceux obtenus dans des spectres de sédiments lacustres actuels, et associés au maximum de représentation du *Podocarpus*. L'existence au cours du Plio-Pléistocène d'une rivière alimentant en rive Est le paléolac explique la présence de ces éléments montagnards dans nos spectres fossiles. Cependant, leurs pourcentages importants témoignent, par comparaison avec l'actuel, d'un drainage important des plateaux entourant, à l'Est et au Nord-Est, le paléolac ou alors d'une extension des forêts plus grande avec rapporchement par rapport au paléorivage. En fait, l'une ou l'autre hypothèse, sont la conséquence d'un même phénomène: des conditions climatiques humides antérieurement à 2.1 MA.

3.1.2. *Période comprise entre 2.1 et 1.8 MA.* Au cours de cette période, on observe, pour 4 spectres, la disparition (*Ilex, Ericaceae*) ou la plus faible représentation (*Myricaceae, Pteridophytes*) des taxa caractéristiques de la période précédente. Ceux-ci sont remplacés par le *Juniperus* en pourcentages supérieurs à ceux trouvés actuellement dans les sédiments lacustres, qui indique en altitude, des conditions climatiques plus sèches mais aussi plus froides.

3.1.3. *Période postérieure à 1.8 MA.* Les tendances froides mises en évidence précédemment s'accentuent au cours de cette période avec un grand développement des forêts à *Juniperus* associé à *Olea*. A nouveau, un plus grand apport de taxa de forêts et de *Pteridophytes* laisse supposer que les conditions climatiques redeviennent plus humides, mais moins que lors de la période antérieure à 2.1 MA.

Schématiquement, nous avons donc la succession climatique suivante:
— entre 2.7 et 2.1 MA: climat humide et chaud,
— entre 2.1 et 1.8 MA: climat sec avec tendances vers un refroidissement,
— entre 1.8 et 1.5 MA: climat plus humide et encore froid.
Pour la période de 2.35 à 1.2 MA, des conclusions identiques avaient été déduites de l'analyse pollinique des sédiments de la basse vallée de l'Omo (Bonnefille 1976, 1979). On retrouve donc les mêmes phénomènes dans deux séquences sédimentaires du même bassin.

De plus, le changement climatique intervenu autour de 1.9 MA, marqué par une importante diminution de la pluviosité, est également confirmé par l'analyse des isotopes de l'oxygène (Cerling *et al.* 1977) et les données paléontologiques (Harris 1976).

3.2. *Dépôts holocènes: Formation de Galana Boi*

Les dépôts holocènes d'Est Turkana (Vondra *et al.* 1971) recouvrent en discordance les formations plio-pléistocènes étudiées précédemment. Ils correspondent à deux hauts niveaux lacustres datés respectivement de 10 000-7 000 BP et 6 200-3 900 BP (Butzer 1980).

Quatre spectres fossiles ont été extraits des sédiments lacustres holocènes. Ceux-ci sont remarquables par leur homogénéité. Ils sont caractérisés par la très grande abondance des *Gramineae* (> 80 %). Les taxa de savane, essentiellement herbacés ne dépassent pas 12 % du total des pollens comptés. Les taxa allochtones de forêt d'altitude ne sont présents que dans un seul spectre, en pourcentage très faible (0.4 %). On trouve également quelques *Typha* témoignant de la présence d'arrivée d'eau fraiche ainsi que de spores de *Pteridophytes.*

L'absence ou la très faible représentation des taxa de forêt d'altitude, le grand développement des *Gramineae* jamais rencontré dans l'analyse de sédiments actuels, ne permettent pas d'interpréter ces végétations en termes climatiques, bien que des oscillations humides aient été définies à cette époque dans d'autres lacs du Rift (Kendall 1969, Butzer 1978). Une analyse plus complète de ces dépôts lacustres, en relation avec les travaux géologiques, sédimentologiques et paléogéographiques en cours de Renaut, Owen & Barthelme permettra une meilleure compréhension et interprétation des résultats.

4. CONCLUSIONS

Les travaux palynologiques menés à l'Est du Lac Turkana dans le cadre d'un vaste programme multidisciplinaire, contribuent d'une façon unique aux reconstitutions de l'environnement du bassin du lac Turkana au cours du Plioquaternaire.

Les résultats fossiles obtenus sont interprétés sur la base de données actuelles qui ont permis d'évaluer quantitativement et qualitativement le transport des pollens par voie aérienne et fluviatile et de mettre en évidence l'influence des conditions de sédimentation sur leur distribution en milieu lacustre.

Confrontés aux conclusions apportées par d'autres études, les données palynologiques fossiles permettent de confirmer, mais aussi de compléter et préciser, les résultats obtenus sur le paléoenvironnement végétal et climatique du Lac Turkana en particulier pour la période comprise entre 2.5 et 1.5 MA.

BIBLIOGRAPHIE

Bonnefille, R. 1972. *Associations polliniques actuelles et quaternaires en Ethiopie (vallées de l'Awash et de l'Omo).* Thèse Doctorat d'Etat, Université de Paris VI, 513pp.

Bonnefille, R. 1976. Palynological evidence for an important change in the vegetation of the Omo Basin between 2.5 and 2 million years ago. In: Y.Coppens *et al.* (eds.), *Earliest man and environment in the Lake Rudolf Basin*: 421-431. University of Chicago Press.

Bonnefille, R. 1976. Implications of pollen assemblage from the Koobi Fora Formation, East Rudolf, Kenya. *Nature* 264: 403-407.

Bonnefille, R. 1979. Méthode palynologique et reconstitutions paléoclimatiques au Cénozoïque dans le Rift Est Africain. *Bull. Soc. Géol. France* (7)21(3): 331-342.

Bonnefille, R. & A.Vincens 1977. Représentation pollinique d'environnements arides à l'Est du lac Turkana (Kenya). In: *Recherches françaises sur le Quaternaire.* INQUA 1977, suppl. au Bull. de l'AFEQ, 50: 235-247.

Bonnefille, R. & A.Vincens sous presse. Analyses palynologiques des sédiments pliopléistocènes situés sous le tuf KBS, Lac Turkana (Kenya). *Actes du VIIIe Congr. Panafr. Préhist. et d'Et. du Quatern.,* Nairobi, Sept. 1977, 5pp.

Bonnefille, R., F.Gasse, C.Azema & M.Denefle sous presse. Palynologie et interprétation paléoclimatique de trois niveaux Pléistocène supérieur d'un sondage du Lac Abhé (Afar, Territoire de Djibouti). *Mém. Muséum Nat. d'Hist. Nat.* Série B, Botanique, tome 27.

Bonny, A.P. 1978. The effect of pollen recruitment processes on pollen distribution over the sediment surface of a small lake in Cumbria. *J. of Ecology* 66: 385-416.

Bonny, A.P. 1980. Seasonal and annual variation over 5 years in contemporary airborne pollen trapped at a Cumbrian lake. *J. of Ecology* 68(2): 421-442.

Butzer, K.W. 1978. Climate patterns in an un-glaciated continent. *Geograph. Magazine* 51(3): 201-208.

Butzer, K.W. 1980. The Holocene Lake Plain of North Rudolf, East Africa. *Physical Geography* 1(1): 42-58.

Cerling, T.E., R.L.Hay & J.R.O'Neil 1977. Isotopic evidence for dramatic climatic changes in East Africa during Pleistocene. *Nature* 267: 137-138.

Davis, M.B. 1967. Pollen deposition in lakes as measured by sediment traps. *Geol. Soc. of Americ. Bull.* 78: 849-858.

Davis, M.B. 1968. Pollen grains in lake sediments: Redeposition caused by seasonal water circulation. *Science* 162: 796-799.

Davis, M.B., L.B.Brubaker & J.M.Beiswenger 1971. Pollen grains in lake sediments: Pollen percentages in surface sediments from Southern Michigan. *Quaternary Research* 1(4): 450-467.

Davis, R.B., L.A.Brewster & J.Sutherland 1969. Variation in pollen spectra within lakes. *Pollen et Spores* 9(3): 558-620.

Hamilton, A.C. 1972. The interpretation of pollen diagrams from Highland Uganda. *Palaeoecology of Africa* 7: 45-149.

Harris, J.M. 1976. Bovidae from the East Rudolf succession. In: Coppens *et al.* (eds.), *Earliest Man and Environments in the Lake Rudolf Basin.* Univ. Chicago Press: 293-301.

Horowitz, A. 1969. Recent pollen sedimentation in Lake Kinneret, Israel. *Pollen et Spores* 11(2): 253-384.

Kendall, R.L. 1969. An ecological history of the Lake Victoria Basin. *Ecol. Monogr.* 39: 121-176.

Leakey, M.G. & R.E.F.Leakey (eds.) 1978. *Koobi Fora Research Project,* vol.1: The fossil hominids and an introduction to their context, 1968-1974. Clarendon Press, Oxford, 250pp.

Maley, J. 1972. La sédimentation pollinique actuelle dans la zone du lac Tchad (Afrique centrale). *Pollen et Spores* 14(3): 263-307.

Maley, J. 1980. *Etudes palynologiques dans le bassin du Tchad et paléoclimatologie de l'Afrique Nord Tropicale de 30 000 ans à l'époque actuelle.* Thèse Doctorat d'Etat, Université de Montpellier, 586pp.

Rossignol, M. 1969. Sédimentation palynologique récente dans la mer Morte. *Pollen et Spores* 11(1): 17-38.

Vincens, A. 1979. Analyse palynologique du site archéologique FxJj50, Formation de Koobi Fora, Est Turkana (Kenya). *Bull. Soc. géol. France* (7)21(3): 443-447.

Vincens, A. 1979. Sédimentation pollinique actuelles dans le bassin du lac Turkana: un modèle pour l'interprétation des données plio-pléistocènes. *Bull. Ass. Sénégal. Et. Quatern. Afr.* 56-57: 74-75.

Vincens, A. sous presse. Interprétation climatique des données palynologiques plio-pléistocènes dans la région Est du lac Turkana (Kenya). *Mém. Muséum Nat. d'Hist. Nat.,* Série B, Botanique, 27, 11pp.

Vondra, C.F., G.D.Johnson, B.E.Brown & A.K.Behrensmeyer 1971. Preliminary stratigraphical studies of the East Rudolf Basin, Kenya. *Nature* 231: 245-248.

ETUDE PALYNOLOGIQUE DES SEDIMENTS
QUATERNAIRES DU LAC ABIYATA (ETHIOPIE)

A. M. LEZINE

Laboratoire de Géologie du Quaternaire, CNSR, Luminy, Marseille, France

SUMMARY

Fossil pollen of the upper 30 m of a core with a length of 162 m taken near the shore of Lake Abiyata in the Ethiopian Rift Valley has been analysed. The results are compared with surface pollen spectra of different vegetation types. The sequence covers the time span of approximately 40 000 to 6 000 years BP. The pollen spectra give indications of semi-arid and more humid periods while colder intervals may have occurred. The results are correlated with changes in lake levels.

INTRODUCTION

Les recherches palynologiques présentées ici s'insèrent dans une étude multi-disciplinaire entreprise par R.Bonnefille, F.Gasse et F.A.Street sur le sondage du lac Abiyata afin de reconstituer les paléoclimats et paléoenvironnements successifs du Quaternaire récent de cette région de l'Ethiopie.* Ce sondage, de 162 m de long, avait été effectué par le Ministère des Mines d'Ethiopie en 1974 dans un but de prospection de gites salifères.

L'analyse palynologique des niveaux supérieurs du sondage fait l'objet d'une thèse de 3è cycle et complète les études sédimentologiques (F.A.Street 1979) et limnologiques (F.Gasse & C.Descourtieux 1979) déjà entreprises. Elle présente un intérêt particulier du fait de la localisation géographique du sondage (cf. carte): il est situé à 1 578 m d'altitude dans le Rift par 7°42' de latitude Nord et 38°36' de longitude Est, dans une région recevant actuellement 550 mm de pluie par an occupée par une végétation de fourré à *Acacia*. On observe sur l'escarpement oriental du Rift situé à une trentaine de kilomètres, un étagement de zones de végétation depuis le fourré à *Acacia* jusqu'à l'étage afro-alpin au-dessus de 3 000 m d'altitude selon un gradient pluviométrique crois-

---

* Etudes réalisées dans le cadre de l'ATP internationale R.Bonnefille 20-43 du CNRS, France.

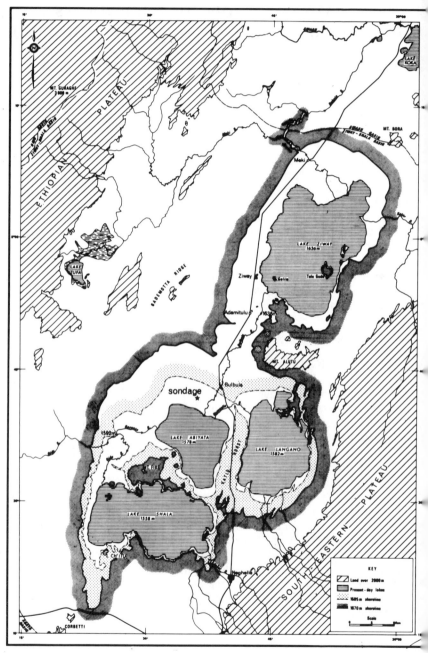

Figure 1. Bassin du Lac Abiyata (Ethiopie).

94

sant et thermique inverse. Les formations végétales intermédiaires sont la savane à Combretacées, la forêt à *Podocarpus gracilior* et *Croton machrostachys, Podocarpus gracilior* et *Olea hochstetteri* puis *Podocarpus gracilior, Juniperus procera* et *Ilex mitis*. Cette étude comporte un intérêt supplémentaire car elle apporte d'importants compléments aux connaissances encore peu nombreuses acquises sur les sédiments lacustres du Quaternaire récent localisés dans les zones de moyenne altitude d'Afrique de l'Est.

L'interprétation paléoclimatique des données palynologiques a été menée sur la base de l'observation des faits botaniques et palynologiques actuels.

1. ÉTUDE DES RETOMBÉES POLLINIQUES ACTUELLES DANS LE BASSIN ZWAY-SHALA (table)

Nous avons dans un premier temps, entrepris l'étude des retombées polliniques actuelles aux abords des lacs en secteur semi-aride et sur l'escarpement oriental du Rift actuellement recouvert par la forêt humide de montagne (forêt de Shashemane).

Cette étude, encore préliminaire, a porté sur 11 échantillons de sol et de vase prélevés dans les lacs et rivières voisins (lac Zway, Langano, Shala et rivière Bulbula), à proximité du lac Shala et le long d'un transect traversant sur 700 m de dénivellation la forêt de Shashemane, de Shashemane à Kofole, par R.Bonnefille et G.Riollet en 1974. Elle a permis de déterminer, d'une part l'image pollinique des différentes associations végétales existant actuellement dans cette région du Rift et répondant à des exigences écologiques connues, d'autre part l'aptitude au transport sur de longues distances des pollens qui leur sont associés.

Dans les échantillons de surface de la région du lac, nous observons que les pollens des plantes du secteur aride (fourré à *Acacia* et savane à *Combretum terminalia*) dont les valeurs sont de 40 % dans les échantillons de sol situés au sein de la formation n'atteignent que 3 % dans les échantillons de vase de lac du même endroit. La pollinisation de ces plantes s'effectue donc de façon essentiellement locale.

Dans les échantillons de surface prélevés sur le transect altitudinal, le transport pollinique sur de longues distances affecte principalement les taxa rapportés à la forêt humide de montagne, notamment *Podocarpus, Olea* sp., *Juniperus* et *Hagenia*. D'autres taxa, tels que *Croton, Araliaceae* sp. et *Ilex*, appartiennent à la même formation, ne se transportent pas ou en très petite quantité. Dans le secteur que nous avons étudié, les pourcentages polliniques de la forêt de montagne sont égaux sous couvert forestier à 50-70 % du total des pollens comptés et diminuent rapidement pour n'atteindre que des valeurs inférieures à 10 à une distance d'une trentaine de kilomètres, ceci malgré la direction favorable des vents dominants.

Tableau 1. Valeurs de la pluie pollinique actuelle dans la région du lac Abiyata (Ethiopie).

Echantillons	Dépression Zway-Shala					Transect Shashemane-Kofole				
Provenance des pollens	Lac Zway vase	riv. Bulbula vase	Lac Shala vase	Shala 36 sol	Shala 33 sol	No.1	No.2	No.3	No.4	No.5
Secteur humide-semi-humide	0,64 %	6 %	18 %	22 %	24 %	74 %	51 %	49 %	29 %	18 %
Secteur semi-humide-semi-aride	1	28	3	40	26	1	4	2	2	2
Hydrophytes	0,42	8	1	3	2	1	3	5	1	1
Ubiquistes	97	50	71	27	40	11	17	27	50	66
Total	2 650	535	525	450	705	396	549	499	564	538

2. ETUDE PALYNOLOGIQUE DU SONDAGE DU LAC ABIYATA

Nous servant, à titre indicatif, des résultats acquis sur la distribution actuelle des retombées polliniques, nous avons entrepris l'analyse des sédiments du sondage et leur interprétation en terme de paléoclimats. Les trente premiers mètres du sondage ont été étudiés, donnant pour une période approximativement comprise entre 40 000 et 6 000 ans BP la succession de plusieurs épisodes climatiques.

2.1. *Période antérieure à 30 000 ans BP (comprise entre 30 m et 19,70 m)*

Pour cette période, l'observation des données palynologiques permet de déterminer une tendance climatique générale semi-aride, proche de l'actuelle, avec des variations de température peu marquées: Deux épisodes semi-arides encadrent une séquence de température nettement inférieure mise en évidence par l'apparition dans les spectres polliniques d'éléments froids de la végétation (*Artemisia, Ericaceae*), elle même suivie d'une période d'intense ruissellement témoignée par l'abondance des spores de Bryophytes et de Ptéridophytes. La fin de la période est marquée par l'accroissement de la végétation de montagne, témoin d'une augmentation de l'humidité.

2.2. *Période comprise entre 30 000(?) et 10 000(?) ans BP (19-8 m)*

La discontinuité de la sédimentation pollinique de cette séquence ne permet pas d'apporter de détails précis quant à la succession climatique de cette longue période de temps. Toutefois, la très forte représentation des taxa de montagne autour de 10 m de profondeur (18 à 31 % du total des pollens comptés) témoigne de l'importante humidité qui à accompagné l'épisode Ziway-Shala III de F.A.Street (1979).

2.3. *La fin du Pléistocène supérieur et l'Holocène ancien (8 m-sommet)*

A l'extrème fin du Pléistocène supérieur et à l'Holocène ancien, cinq épisodes, aux caractéristiques climatiques nettement différentes, ont été mis en évidence. Un premier épisode d'augmentation progressive de l'humidité entraine la remontée irrégulière du niveau lacustre jusqu'à un niveau peu élevé, proche du lieu de sondage (8-´5 m). L'arrêt de cette remontée lacustre est indiqué entre 5,5 m et 5 m de profondeur par la présence d'un sol et par l'importance des pollens de plantes hydrophiles dans les spectres polliniques: les pollens de *Typha* atteignent ici les pourcentages exceptionnel de 38 %. Cet épisode de stabilisation est daté dans le sondage de 9 950 ± 170 BP (Gif 3968). Entre 5 m et 1,70 m de profondeur s'individualise un troisième épisode caractérisé par un exceptionnel accroissement de la représentation des éléments froids de la végétation. Ils atteignent en effet jusqu'à 7 % du total des pollens comptés, ce qui est sans rapport avec ce que nous pouvons observer actuellement dans la représentation pollinique de la végétation actuelle de la dépression ou du secteur forestier. Ces pourcentages témoignent soit d'un refroidissement climatique de grande ampleur soit d'un régime des pluies différent de l'actuel avec des pluies mieux réparties au cours de l'année. Cet épisode correspond à un haut niveau lacustre durant lequel le paléolac Zway-Shala V, ayant atteint l'altitude maximale de 1 673 m aurait déversé le trop plein de ses eaux dans le bassin de l'Awash situé plus au Nord. Une période de très grande aridité lui fait suite vers 7 000 BP, témoignée par le remaniement des sédiments, la corrosion des pollens et des diatomées entre 1,70 et 1 m de profondeur. Un dernier épisode d'humidité exceptionnelle voit le développement de la forêt de montagne à *Podocarpus* et son déplacement vers de plus basses altitudes qu'actuellement à proximité du lac. Il correspond, selon toute vraisemblance au haut niveau lacustre Ziway-Shala VI.

CONCLUSION

Les études multidisciplinaires effectuées sur les sédiments du lac Abiyata ont mis en évidence un certain nombre de fluctuation des niveaux lacustres. L'analyse des pollens, en retraçant les paléoenvironnements végétaux successifs de l'ensemble du bassin à permis de préciser l'origine climatique de ces fluctuations, à la suite d'une interprétation discutée à partir d'observations préliminaires concernant les retombées polliniques actuelles dans l'ensemble du bassin versant.

Ce travail à pu être réalisé grâce à une collaboration entre le Laboratoire du Géologie du Quaternaire (CNRS — Marseille) et l'Institut du Quaternaire (Université de Bordeaux I) dont nous remercions les deux directeurs H.Faure et F.Bordes. Nous remercions également R.Bonnefille qui nous à confié ce matériel d'étude, N. & G.Buchet pour leur collaboration technique.

BIBLIOGRAPHIE

Gasse, F. & C.Descourtieux 1979. Diatomées et évolution de trois milieux éthiopiens d'altitude différente au cours du Quaternaire récent. *Palaeoecology of Africa 11:* 117-134.
Gasse, F. & F.A.Street 1978. Late Quaternary lake-level fluctuations and environments of the northern Rift Valley and Afar region (Ethiopia and Djibouti). *Palaeogeogr., Palaeoclimatol., Palaeoecol.* 24: 279-325.
Grove, A.T., F.A.Street & A.S.Goudie 1975. Former lake levels and climatic change in the Rift Valley of Southern Ethiopia. *Geogr. J.* 141: 177-202.
Lezine, A.M. & R.Bonnefille 1980. Séquence pollinique et paléoenvironnements du lac Abiyata (Ethiopia) au Pléistocène supérieur et à l'Holocène. 8e Réun. Ann. Sci. Terre, Marseille, Févr. 1980, 227, Paris: Soc. Géol. Fr.
Street, F.A. 1979. *Late Quaternary lakes in the Ziway-Shala basin.* PhD Thesis unpubl. Univ. of Cambridge, 457pp.

TWENTIETH CENTURY FLUCTUATIONS IN LAKE LEVEL IN THE ZIWAY-SHALA BASIN, ETHIOPIA

F. A. STREET-PERROTT

School of Geography, Oxford, UK

ABSTRACT

This note summarises the historical evidence for fluctuations in the level of the closed lakes Abiyata and Shala during the 20th century. Published staff-gauge data exist only for the period 1967-1974, but can be supplemented by historical reports and photographs dating back to 1926. Low water levels were encountered in (1926), 1933, 1956, 1967 and 1978, and high levels in 1938 and 1970-72. The observed maxima and minima are closely related to rainfall variations in Addis Ababa, but show no correlation with rainfall or lake-level fluctuations in Southwestern Ethiopia.

INTRODUCTION

In recent years, attention has been focussed on historical lake-level fluctuations as an indicator of changes in the atmospheric circulation over Africa (Lamb 1966, Butzer 1971, Vincent *et al.* 1979, Nicholson 1980). Lake Chad and a number of East African lakes have furnished records beginning before 1900 AD. Nicholson (1980) has noted that these show a general pattern of high levels in the late 19th century, followed by a sharp decrease around 1900 and an abrupt recovery in the 1960's. In detail, however, there are marked dissimilarities between individual curves, which can be attributed to differences in water balance and regional climate. Closed lakes such as Turkana and Naivasha are less subject to short term (< 20 y) oscillations, and fluctuate through a larger vertical range, than open lakes like Victoria (Mörth 1967, Butzer 1971, Vincent *et al.* 1979). The lack of an outlet means that changes in water storage are essentially cumulative, thus damping climatic fluctuations of short duration (Street 1980).

Climatic records from Ethiopia covering the period before World War II are very scarce (Fantoli 1966). The longest rainfall series, that from Addis Ababa, dates back only to 1898, and the majority of stations began observations in the early 50's. Most river-discharge series are also relatively short.

99

Table 1. Mean monthly and annual river discharges in the Ziway-Shala Basin during the period 1969 to 1973 (million cubic metres)

River	Jan	Feb	Mar	Apr	May	June	July	Aug	Sept	Oct	Nov	Dec	Total
Katar[1] at Ogelcho	6.69	7.32	19.63	15.76	16.35	12.31	62.17	145.33	90.17	30.07	8.49	6.07	420.36
Meki[1] at Meki Town	6.93	12.93	31.61	31.91	21.61	21.91	78.92	124.52	68.95	23.81	8.12	5.38	436.60
Gedemso[2] near Lake Langano	0.71	1.07	4.47	5.24	4.17	4.49	17.00	20.67	15.06	7.03	1.76	0.78	82.45
Gorgeza[1]	0.69	0.68	1.40	0.49	0.72	0.75	2.15	3.41	2.50	0.81	1.17	0.81	15.58
Bulbula[1]	14.98	10.12	10.16	8.16	7.10	5.40	8.08	24.03	45.84	43.63	20.42	12.93	210.85
Horocallo[1]	3.86	2.53	1.82	1.42	1.23	0.91	1.06	4.80	12.74	15.54	7.39	4.36	57.66

Sources: 1. Makin et al. (1976).
2. Kingham, personal communication 1976 (AVA data, for 1969-72 only).

This paper presents published staff-gauge data for the lakes in the Ziway-Shala basin, covering the period since 1967, and reviews the sporadic information provided by earlier historical reports and photographs extending back to 1926.

GEOGRAPHICAL SETTING

The Ziway-Shala Basin is situated in the Main Ethiopian Rift at about $7°30'$N. At the present day, it contains four large alkaline lakes at elevations ranging from 1 558 to 1 636 m. These form two disjunct drainage systems (figure 1). Lake Ziway is fed by two large rivers: the Katar, which rises above 4 000 m on the Arussi Mountains to the southeast, and the Meki, which drains the western Rift escarpment and foothills. During the rainy season, Lake Ziway overflows southwards into Lake Abiyata via the Bulbula River. Lake Abiyata is at present without outlet. It also receives runoff from the eastern escarpment via Lake Langano and its outflow, the Horocallo River, and from the western escarpment via the Gorgeza channel, which is a rainy season distributary of the Gidu River. The main channel of the Gidu flows southwards into Lake Shala, which again lacks a surface outlet. Lake Shala is also fed by runoff from the Adabar River, derived from the Rift escarpment to the southeast. Abiyata and Shala have remained separate throughout the historical period (Street 1979: ch.5).

The relative importance of the rivers which feed the lakes can be judged in part from the data shown in table 1. These were collected during 1969-1973, which was a relatively wet period. There are no discharge data for the Gidu, the Adabar, or any of the rivers feeding Lake Langano apart from the Gedemso. Street (1979) estimated that direct rainfall onto the lake surfaces amounted to 20-25 % of the total inputs and spring discharge to less than 1 %. The water balance of the basin as a whole is therefore dominated by the Meki and the Katar, which drain the mountain areas in its northeastern half.

All the rivers display a major discharge maximum in July-September and a secondary maximum in March-May (table 1). This pattern reflects the bimodal distribution of precipitation in central Ethiopia (Kingham 1975). Bethke (1976) maps the Ziway-Shala Basin at the southwestern tip of his climatic zone III (Escarpment-Rift). This group includes stations like Addis Ababa, Meqele and Dire Dawa which are situated on the margins of the Afar triangle or in the northern part of the Main Rift. All these stations show a primary rainfall maximum in July-August and a secondary maximum in February-May.

SOURCES OF LAKE-LEVEL DATA

The first report of the existence of lakes in the northern part of the Main Rift dates from the Portuguese colonial period. The missionary Manoel de Al-

101

meida wrote around 1628 that the Haoax [Awash] 'receives the waters of a big river named Machŷ [Meki] coming from Lake Zuaĵ [Ziway]' (Beckingham & Huntingford 1954). This intriguing idea is also embodied in Coronelli's *Atlante Veneto* (1690) (Grove *et al.* 1975, plate VIIa). But since Lake Ziway would have to rise by 35 m in order to overflow into the Awash, the postulated connection should be viewed with some scepticism (Street 1979: 149).

The existence and interrelationships of the other three lakes were not established until the Erlanger expedition (Neumann 1901, 1902). Observations of relevance to recent lake-level fluctuations began with Dr Hugh Scott's expedition in 1926 (Omer-Cooper 1930) and the visit by Eric Nilsson in 1933 (Nilsson 1940). The first bathymetric surveys were carried out by the Italian wartime limnological mission (Vatova 1941) and updated by Italconsult (1970). I have also made use of aerial photographs taken by Hunting Surveys in 1956 and 1972, and by Mr A.T.Grove in 1978. Additional observations of Lake Abiyata were made in 1964 and 1967 by Dr Emil Urban (personal communication) and in 1974 by the UK Land Resources Division (Makin *et al.* 1976). Gauging-station data exist only for the period since 1967 (Ziway) and 1968 (Langano, Abiyata).I have not been able to obtain access to official data covering the period after the revolution in 1974.

FLUCTUATIONS OF LAKE SHALA

The fluctuations of Lake Shala were first noted by Nilsson (1940). He reported finding drowned *Acacia* trees extending up to 8 m above the 1933 water surface and at least 1 m below it, demonstrating that the level of the lake had previously risen by at least 9 m and then fallen again. Nilsson published a photograph (1940: figure 28) showing a substantial number of well preserved trunks up to 60 cm in diameter and several metres in height. The site of the photograph was probably the northwestern shore. Supporting evidence for a recent rise can be found around the eastern margin of the lake, where very fresh Gilbert-type deltas, bouldery strandlines and inactive spring vents testify to a shoreline 5-6 m above the 1974 water surface (Street 1979: 151).

A certain amount of confusion has crept into the literature because it now appears that there is more than one generation of drowned trees around the shore. A sample of wood collected on behalf of H.Smeds (Smeds 1964) was identified as the leguminous shrub ambatch (*Aeschynomene elaphroxylon*). This is immediately surprising, because *Aeschynomene* typically grows on the outer fringes of freshwater swamps. It is common around oligohaline lakes such as Ziway, Abaya, Naivasha, Baringo and Chad, and in the White Nile swamps (Beadle 1932, Brunelli *et al.* 1941, Rzóska 1976). Lake Shala is about 266 m deep, with a dissolved solids content approaching 16 g/l. It would require a large increase in volume at the present day to dilute the waters sufficiently to bring them into the oligohaline category (<2 g/l).

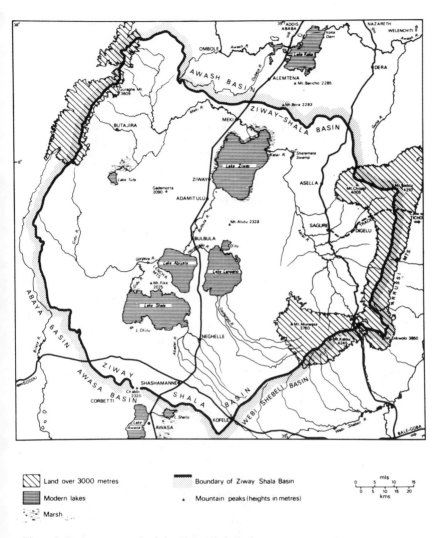

Figure 1. Drainage network of the Ziway-Shala Basin.

On the southeastern shore of the lake, carbonized stumps extend up to 2.5 m above the October 1974 lake level. I collected a sample of the wood which was identified by the Jodrell Laboratory at Kew as *Acacia* sp., probably *Acacia nilotica.* This is a species common on the margins of African lakes, pools and rivers. It can tolerate seasonal flooding for several months each year, and does not imply any marked decrease in salinity (G. Wickens, per-

103

sonal communication). A sample from the same stump gave a ^{14}C date of
9 800 ± 100 BP (HAR-2787). The corresponding $\delta^{13}C$ value was $-26‰$,
which makes the possibility of substantial contamination by dead carbon
appear very remote (R.L.Otlet, personal communication). It is therefore
clear that at least some of the drowned trees are of early Holocene age, dating
from the lowstand between lacustral maxima Ziway-Shala IV (11 500-ca.
10 200 BP) and Ziway-Shala V (9 400-ca.8 000 BP) (Street 1979). Their
stumps have been exhumed from under a cover of mid-Holocene (Ziway-
Shala VI) sediments. The sample referred to by Smeds may well have been
collected from the same level, so that it cannot be used to make any inferences about recent conditions.

All we can safely conclude is that the level of the lake was relatively low
in 1933 and that Nilsson found traces of a rise in lake level which had occurred
in the relatively recent past.

FLUCTUATIONS OF LAKE ABIYATA

The first interesting observation of Lake Abiyata was made by Omer-Cooper
(1930). He found large dead trees lining the Bulbula River where it enters the
lake, and commented 'that within the last few decades the waters had reached
an abnormal level, and that previously its waters had been lower for some
period sufficiently long to allow the soil to become sweet and trees to grow
to moderate dimensions' (1930: 200). Figure 14 in his paper suggests that
the rise in water level which drowned the trees had occurred not very long
before, because although the bark and foliage had rotted away most of the
trees were still standing, with their main branches intact.

In 1938, the Italians measured the depth of the lake and found it to be
14.2 m, which is the greatest ever recorded (Vatova 1941). It corresponds to
a surface elevation of about 1 581.2 m. This figure may be a slight overesti-
mate, however, since a photograph taken at the time (Ibid: tav.55) shows
that the shoreline lay in an almost identical position to 1972, although much
higher than in 1956 and 1978. From 1938 to 1956, there is a long gap without observations. Makin et al. (1976) estimated from the 1956 aerial photo-
graphy that the lake was then very low, standing about 2 m below datum on
the AVA staff gauge, i.e. at ca.1 574 m (figure 2). Large areas of gently sloping
saline grassland were exposed around the shore. Based on observations by Dr
Emil Urban (personal communication) the 1964 level was very similar.

Between 1964 and 1967, the water level receded by 1-2 m. Lake level in
1967 was the lowest ever recorded, ca.1 573 m, equivalent to a maximum
depth of only 6-7 m. In March 1967, the Horocallo River was dry (cf. figure 3).
Beginning in the rainy season of 1967, Abiyata rose dramatically at a rate of
1.9 m/y, submerging a large number of *Acacia tortilis* trees near the Bulbula
outlet (Urban, personal communication). It reached very high levels in 1969-

104

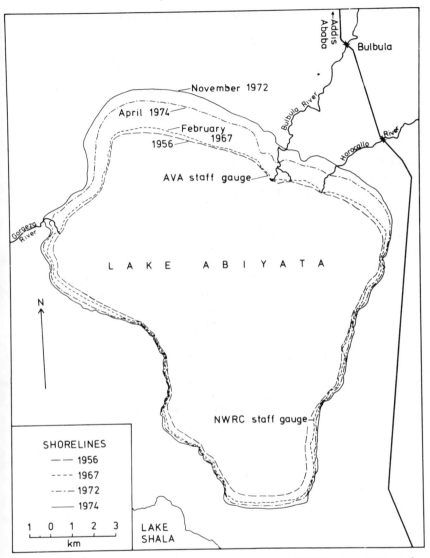

Figure 2. Fluctuations in the extent of Lake Abiyata since 1956 (data from Makin *et al.* 1976).

1973, culminating in a peak of 1 579.4 m in November 1971 (Makin *et al.* 1976, Street 1980: fig.4). By April 1974 the water surface had dropped considerably (figure 2) and between 1974 and 1978 it fell by a further 1-2 m (estimated from an oblique aerial photograph taken by Mr A.T.Grove).

105

FLUCTUATIONS IN RAINFALL AND RUNOFF

Published runoff records show that the fluctuations of Lake Abiyata since 1963 reflect large variations in runoff. The annual discharge of the Meki River rose from a minimum of 121.5 mcm in 1965 to a maximum of 486.4 mcm in 1970 (figure 3). This amounts to a four-fold increase. By 1973, the total had fallen to 362.4 mcm. Likewise, the Katar reached a peak of 681.1 mcm in 1970 in comparison with 261.6 mcm in 1974, and the Gedemso River carried 94.9 mcm in 1970 compared with 66.4 mcm in the latest year for which data are available (1972). The input of the Gorgeza spillway into Lake Abiyata has also fallen from 28.7 mcm in 1969 to 1.2 mcm in 1974.

The two lakes with outlets, Ziway and Langano, oscillated by only 1.97 and 1.67 m respectively between 1967/8 and 1974. Both reached a rainy-season maximum in 1969. Since then, their mean levels have declined by 0.172 m/y and 0.152 m/y respectively. This small variation was nevertheless sufficient to cause a very marked fluctuation in annual discharge in the Bulbula and Horocallo Rivers (figure 3), since outflow is an exponential function of water surface height above a critical threshold (Makin et al. 1976, figs.11, 26). The flow of the two rivers peaked in 1970. By 1973 it had dwindled considerably and in 1974-1976 both of them dried up during at least part of the dry season. Comparison of the fluctuations of Abiyata since 1968 with the Meki River hydrograph shows that the level of the lake lagged only 1-2 y behind the variations in river input into the higher lakes, and about 1 y behind the variations in flow in the Bulbala and Horocallo Rivers.

The rainfall records from the Ziway-Shala Basin are short and of very uneven quality (Kingham 1975). But Makin et al. (1976) found a strong correlation between the Meki River hydrograph and annual rainfall totals in Addis Ababa. Using the Addis Ababa data together with rainfall records for the Katar catchment, McKerchar & Douglas (1974) successfully modelled the fluctuations of Lake Ziway since 1967. They were then able to extend their water-balance model back to 1954 to produce the simulated lake-level curve for Ziway shown in figure 4 (curve a). This curve agrees surprisingly well with the recorded fluctuations of Lake Abiyata, allowing for a lag of about 2 y.

Prior to 1954, the historical data from Abiyata and Shala can only be compared with the rainfall curve for Addis Ababa (figure 3). It may be seen that rainfall totals were dropping rapidly in 1926, reaching a minimum in 1932 just before Nilsson's visit to Lake Shala. It is therefore not improbable that the drowned trees photographed by Nilsson and Omer-Cooper dated from lake-level highstands related to the recorded rainfall maxima in 1916 and 1925.

Despite the good correlation between the levels of Lake Abiyata and rainfall totals in Addis Ababa, which also lies in Bethke's zone III (Bethke 1976), there is very little correspondence between the fluctuations of Abiyata and the next lake basin to the south, Awasa (Kingham in Grove et al. 1975), Awasa Town is situated near the northeastern boundary of Bethke's climatic

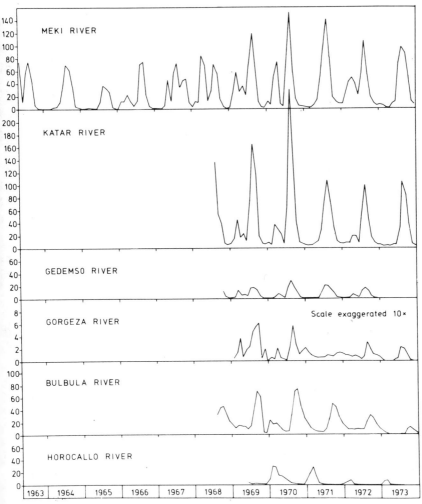

Figure 3. Hydrographs of the rivers in the Ziway-Shala Basin, 1963-1973 (data from Makin *et al.* 1976).

zone I (Southwest), which is a high rainfall area characterized by a single rainfall maximum in April-October. The existence of an important climatic gradient between the Ziway-Shala and Awasa catchments is reflected in the lack of correlation between the annual rainfall series from the two areas (Kingham 1975). However, the fluctuations of Lake Awasa (1 680 m asl) do not correlate well even with local rainfall totals, and appear to have been more strongly influenced by variations in evaporation and the number of days per year with

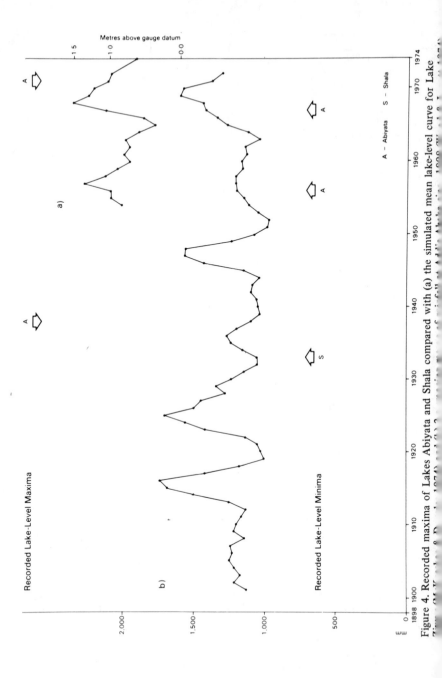

Figure 4. Recorded maxima of Lakes Abiyata and Shala compared with (a) the simulated mean lake-level curve for Lake

108

cloud and rain (Kingham in Grove *et al.* 1975). Comparison of figure 4 with the long discharge series for the Sobat River (Butzer 1971:fig.5-4) confirms that fluctuations in rainfall in the Ziway-Shala Basin were not in phase with climatic zone I.

CONCLUSIONS

The historical data reviewed in this paper indicate that the levels of Lakes Abiyata and Shala have varied by at least 6 and 9 m respectively during the 20th century. These fluctuations are similar in amplitude to those of other closed lakes such as Turkana and Naivasha (Butzer 1971, Vincent *et al.* 1979). Modern gauging data are sufficient to show that the level of Lake Abiyata is closely related to fluctuations in river discharge, allowing for a time lag of about 1-2 y. There is no reliable data to test whether year-to-year changes in cloudiness and evaporation have also played a significant role in the water balance of the lake.

On a slightly longer time scale, the levels of Lakes Shala and Abiyata show a strong correlation with rainfall totals in climatic zone III (Bethke 1976), as exemplified by the rainfall curve for Addis Ababa. There is a strong cyclical tendency in the Addis Ababa record which is at least partially reflected in the lake-level data. Low water levels were encountered in (1926), 1933, 1956, 1967 and 1978, and high levels in 1938 and 1970-72. No information is available to check this relationship between 1938 and 1956. Although the oscillations in rainfall totals in Addis Ababa have been correlated with the 11 y sunspot cycle (Wood & Lovett 1974) it would be premature to regard the lake-level data as confirmatory evidence. Furthermore, the available records are too short to determine whether water levels have also varied significantly on a secular timescale as observed in East Africa.

The lack of similarity between the fluctuations of Abiyata and Awasa, only 65 km to the south, show that it would be unwise to regard the short-term fluctuations experienced in the Ziway-Shala Basin as representative of more than the Rift-Escarpment region of central Ethiopia.

ACKNOWLEDGEMENTS

I thank Mr R.L.Otlet for the radiocarbon determination, and Mr A.T.Grove, Mr T.J.Kingham and Dr Emil Urban for supplying me with data and helpful advice.

REFERENCES

Beadle, L.C. 1932. Scientific results of the Cambridge Expedition to the East African Lakes 1930-31. 4. The waters of some East African lakes in relation to their fauna and flora. *J. Linn. Soc. (Zool.)* 38: 157-211.

Beckingham, C.F. & G.W.B.Huntingford 1954. *Some records of Ethiopia, 1593-1646.* Hakluyt Society, London, 267pp.

Bethke, S. 1976. Basin zonal rainfall patterns in Ethiopia. In: A.M.Hussein (ed.), *Rehab: Drought and famine in Ethiopia.* African Environment: Special Report 2, International African Institute, London: 97-103.

Brunelli, G., G.Cannicci, S.Loffredo, C.M.Maldura, G.Morandini, P.Parenzani, A.Vatova, & G.Zolezzi 1941. *Esplorazione dei Laghi della Fossa Galla: Missione Ittiologica dell'Africa Orientale Italiana,* vol.1. Ministero dell'Africa Italiana, Rome, 258pp.

Butzer, K.W. 1971. *Recent history of an Ethiopian delta: the Omo River and the level of Lake Rudolf.* Univ. Chicago Dept. Geogr. Res. Paper 136: 184pp.

Fantoli, A. (ed.) 1966. *Contributo alla Climatologie dell'Etiopia: riassunto dei risultata e tabell meteorologiche e pluviometriche.* Ministero degli Affari Esteri, Rome, 558pp.

Grove, A.T., F.A.Street & A.S.Goudie 1975. Former lake levels and climatic change in the Rift Valley of southern Ethiopia. *Geogr. J.* 141: 177-202.

Italconsult 1970. *Meki River Diversion Scheme.* Produced for the Imperial Ethiopian Government, Addis Ababa. Rome, 5 vols.

Kingham, T.J. 1975. *Rainfall records for the southern Rift Valley of Ethiopia.* Supplem. Rept. Land Resources Divn., UK Min. Overseas Devel., Tolworth 18: 50pp.

Lamb, H.H. 1966. Climate in the 1960's. Changes in the world's wind circulation reflected in prevailing temperatures, rainfall patterns, and the level of the African lakes. *Geogr. J.* 132: 183-212.

McKerchar, A.I. & J.R.Douglas 1974. *Lake Zwai Study.* Unpubl. Rept., Institute of Hydrology, Wallingford, 65pp.

Makin, M.J., T.J.Kingham, A.E.Waddams, C.J.Birchall & B.W.Eavis 1976. *Prospects for irrigation development around Lake Zwai, Ethiopia.* Land Resources Study, Land Resources Divn., UK Min. Overseas Devel., Tolworth 26: 270pp.

Mörth, H.T. 1967. Investigations into the meteorological aspects of the variations in the level of Lake Victoria. *East African Met. Dept. Mem.* 4(2): 6pp.

Neumann, O. 1901. The Erlanger expedition in north-east Africa. *Geogr. J.* 17: 528-529.

Neumann, O. 1902. From the Somali coast through southern Ethiopia to the Sudan. *Geogr. J.* 20: 373-401.

Nicholson, S.E. 1980. Sahara climates in historic times. In: M.A.J.Williams & H.Faure (eds), *The Sahara and the Nile* Balkema, Rotterdam: 173-200.

Nilsson, E. 1940. Ancient changes of climate in British East Africa and Abyssinia. *Geogr. Annlr.* 22: 1-78.

Omer-Cooper, J. 1930. Dr Hugh Scott's Expedition to Abyssinia – A preliminary investigation of the freshwater fauna of Abyssinia. *Proc. Zool. Soc. Lond.*:195-206.

Rzóska, J. (ed.) 1976. *The Nile: Biology of an ancient river.* Dr W.Junk, The Hague, 417pp.

Smeds, H. 1964. A note on recent volcanic activity on the Ethiopian Plateau, as witnessed by a rise of the level of Lake Wonchi 1400 ± 140 BP. *Acta geogr., Helsinki* 18, 1-32.

Street, F.A. 1979. *Late Quaternary lakes in the Ziway-Shala Basin, southern Ethiopia.* PhD dissertation, Cambridge University, 493pp.

Street, F.A. 1980. The relative importance of climate and local hydrogeological factors in influencing lake-level fluctuations. *Palaeoecology of Africa* 12: 137-158.

Vatova, A. 1941. Relazione sui risultati idrografici relativi ai laghi dell'Africa Orientale Italiana esplorati dalla Missione Ittiologica. In: G.Brunelli *et al.* (eds.), *Esplorazione dei Laghi della Fossa Galla: Missione Ittiologica dell'Africa Orientale Italiana* vol.1. Ministero dell'Africa Italiana, Rome: 67-127.

Vincent, C.E., T.D.Davies & A.K.C.Beresford 1979. Recent changes in the level of Lake Naivasha, Kenya, as an indicator of equatorial westerlies over East Africa. *Climatic Change* 2: 175-189.

Wood, C.A. & R.R.Lovett 1974. Rainfall, drought and the solar cycle. *Nature* 251: 594-596.

MISE EN ÉVIDENCE D'UN NIVEAU MARIN HOLOCÈNE SUBMERGÉ DANS L'ESTUAIRE DU GABON

P. WEYDERT & J. C. ROSSO

Laboratoire de Géologie du Quaternaire, Luminy, 13288 Marseille, France

RÉSUMÉ

Lors de travaux d'approfondissement du Port à Bois d'Owendo, situé sur la rive N de l'estuaire du Gabon (0°17'N, 9°31'E), il a été découvert, sous 4 mètres de vase actuelle, un niveau fossilifère à Mollusques et Madréporaires, témoins de la transgression holocène.

L'analyse biocoenotique des espèces récoltées permet de préciser les conditions qui ont régné au Quaternaire récent dans ce secteur de l'Afrique équatoriale.

1. CADRE GÉOLOGIQUE GÉNERAL

L'estuaire du Gabon, long de 80 km, s'enfonce largement dans la partie septentrionale du bassin côtier gabonais, qui lui-même appartient à une grande unité structurale bordant le côté Ouest du craton congolais (Houreg *et al.* 1954, Hudeley 1970).

Ce bassin est formé très schématiquement d'une série continentale, dite 'antésalifère', et d'une série marine, 'post-salifère' – la période 'salifère' (Aptien supérieur) correspondant on fait au passage d'un grand rift continental. Le sel est, par ailleurs, à l'origine d'importants phénomènes d'halocinèse liés à des mouvements cénomaniens et sénoniens.

L'estuaire du Gabon recoupe sur son parcours toute la série post-salifère pour atteindre, à Kango, les formations néocomiennes (anté-salifères) qui constituent la base de la 'série du Cocobeach' (Hudeley 1970). Il est limité au Sud par une flèche sableuse, la pointe Pongara, dont l'édification, probablement récente, est en relation avec le courant de dérive littorale orienté vers le Nord.

Une vallée sous-marine, utilisée comme chenal de navigation, entaille l'estuaire en diagonale, depuis la pointe d'Owendo jusqu'à la côte (figure 2). Toutefois, le substratum sénonien et turonien, affecté d'un léger pendage, détermine un certain nombre de hauts-fonds, dont l'alignement (NNE-SSW) correspond à la direction des affleurements observés à terre.

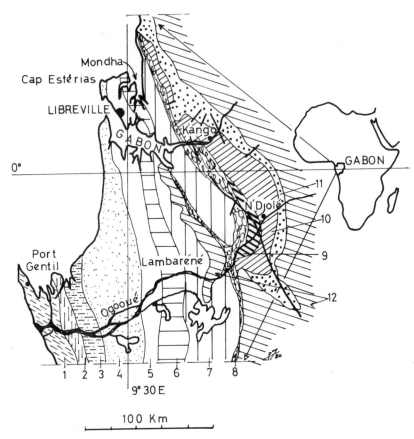

Figure 1. Esquisse géologique de la partie nord du bassin côtier gabonais.
Geological sketch of the northern part of the coastal basin of Gabon.
1 – Miocène; 2 – Eocène; 3 – Paléocène; 4 – Sénonien; 5 – Turonien; 6 – Cénomanien;
7 – Madiéla (Albien); 8-10 – Cocobeach supérieur, moyen et inférieur (Néocomien à
Barrémien); 11 – Précocobeach (Permocarbonifère, Trias, Jurassique); 12 – Socle (Anté-
cambrien).

De récents travaux d'aménagement effectués au Port à Bois, dans la baie
d'Eggoume (S de Libreville) ont fourni l'occasion d'une campagne de sondages
et de dragages, destinés notamment à approfondir le bassin pour atteindre la
cote -4 ou -5 m sous le zéro hydrographique. Compte-tenu du fluage des vases
récentes, l'affouillement a été poussé jusqu'à -7, -8 m. Or les sédiments rejetés
par la drague (vase noire parfois très compacte, graviers, blocs de latérite) ont
livré un important matériel paléontologique (coquilles de Mollusques, Madréporaires) dont nous entreprenons ici l'étude.

112

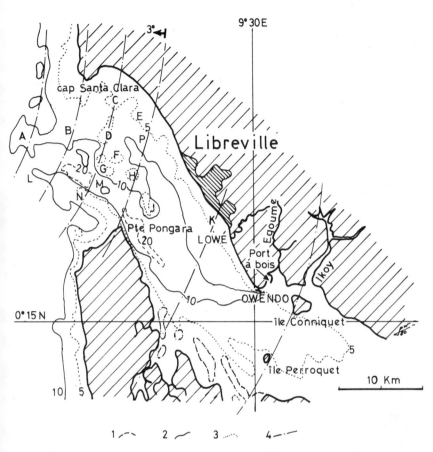

Figure 2. Carte de situation générale de l'estuaire du Gabon.
General situation map of the estuary of Gabon.
A – banc de la Thémis; B – banc du Postillon; C – banc de la Recherche; D – banc du Nisus; E – banc de l'Adour; F – rochers de Vialètes; G – banc de Caraïbe; H – banc du milieu; J – banc du Sud-Est; K – banc de la Malouine; L – banc de la Mouche; M – banc du Papillon; N – banc de Pongara; P – banc du Caïman.
1 – isobath-20; 2 – isobath-10; 3 – isobath-5; 4 – alignements d'affleurements rocheux (aligments of rocky outcrops).

2. DESCRIPTION DU REMPLISSAGE DE LA BALE D'EGOUME

Au cours de la phase préliminaire à l'aménagement du port, 18 sondages ont été pratiqués sur le rivage Ouest de la baie d'Egoume (figure 4 et 5). Il est donc possible de reconstituer, au moins partiellement, la séquence sédimentaire (Dwars 1967).

Tableau 1. Liste systématique des Mollusques récoltés.

Liste des espèces du Port à Bois d'Owendo	Espèces existant dans les sables de la Lowé	Espèces indicatrices du milieu marin franc	Espèces indicatrices des milieux laguno vaseux abrités	Espèces indicatrices des milieux dessalés
GASTEROPODES				
Archimediella annulata (Kiener)	•	•		
Archimediella gemmata (Reeve)	•	•		
Bulla striata adansoni (Philippi)	•	•	•	•
Genota mitraeformis (Wood)	•	•		
Hexaplex rosarium (Röding)		•		
Hinia tritoniformis (Kiener)	•	•		
Natica marochiensis (Gmelin)			•	
Nerita senegalensis (Gmelin)		•		
Ocenebra inermicosta (E.M. Vokes)		•		
Pugilina morio (L.)			•	
Thaïs (Thaisella) callifera callifera (Lamarck)			•	
Turritella ungulina (L.)	•	•		
Thaïs (Stramonita) forbesi (Dunker)			•	
Tympanotonus fuscatus (L.)	•			•
Tympanotonus fuscatus radula (L.)				•
SCAPHOPODES				
Dentalium (Antalis) katchekense (Fischer-Nicklĕs)	•	•	•	
LAMELLIBRANCHES				
Aequipecten flabellum (Gmelin)	•		•	
Anadara (Senilia) senilis (L.)	•		•	•
Arcopsis (Striarca) lactea (L.)		•		
Clinocardium kobelti (von Maltzan)	•	•		
Crassostrea gasar (Dautzenberg)			•	•
Corbula (Cuneocorbula) dautzenbergi (Lamy)		•		
Dosinia (Sinodiella) isocardia (Dunker)			•	
Dosinia (Asa) sp.			•	
Felania diaphana (Gmelin)			•	
Iphigenia laevigata (Gmelin)	•		•	•
Nuculana (Lembulus) bicuspidata (Gould)	•	•	•	
Ostrea (Crassostrea) sp.				
Pseudochama gryphina (Lamarck)		•		
Tellina (Eurytellina) senegambiensis (Salisbury)		•		
MADREPORAIRES				
Schizoculina fissipara (ME et M)		•		

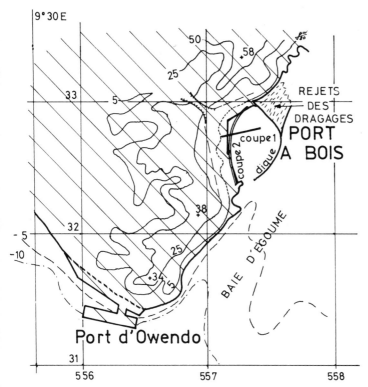

Figure 3. Carte de situation du Port à Bois d'Owendo.
Situation map of the Port à Bois d'Owendo.

2.1. *La coupe 1* (figure 4) correspond à l'ancienne ligne de rivage, avant agrandissement du bassin portuaire. On remarque que le substratum rocheux est aux environs de 5 m du côté S, avec une remontée à 1,2 m, et qu'on le retrouve, côté N, à 2,8 m. Entre les sondages D et F, il n'a pas été atteint. On constate également une grande diversité de corps sédimentaires, aux faciès variés, dont l'éventail ne fait que reproduire celui du biseau littoral, entaillé par un versant de côte relativement abrupt (voir figure 5, coupe 2). Des accumulations de latérite, probablement remaniées, se rencontrent localement.

2.2. *La coupe 2* (figure 5), orientée E-W dans l'axe du port à Bois, permet de préciser le remplissage de la paléovallée de l'Egoume. Le profil du substratum rocheux est, comme on le voit, assez 'raide'. Vers le large, les sondages se sont arrêtés sur de la latérite à des cotes variant de 22,7 à 26,1 m. On peut présumer que cette latérite s'est accumulée au pied du versant lors d'une phase régressive.

115

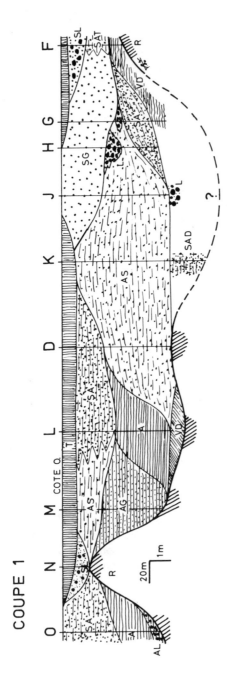

Figure 4. Coupe schématique selon la cote zéro (voir fig.3 pour la situation). Schematic cross section along the zero level (see fig.3 for the location). A – argile (clay); AG – argile grise (grey clay); AL – argile et latérite (clay with laterite rock); AS – argile sableuse (sandy clay); R – roche (rock); SA – sable argileux (clayey sand); SAD – sable et argile dure (sand with compact clay); SAT – sable argileux et tourbe (clayey sand and peat); SG – sable gris (grey sand); SL – sable et gravier de latérite (sand with laterite gravel); T – peat; VD – compact mud.

116

Jusqu'à la cote -4 m, la paléovallée est remplié d'une vingtaine de mètres de vase noire, compacte, tourbeuse par endroits. Celle-ci contient, au sommet, des blocs et des graviers de latérite, ainsi que des blocs de calcaire.

Dans les rapports de sondage, il est fait mention de deux niveaux conquilliers: le premier vers 18 m, le second vers 7,5 m. C'est de ce dernier niveau que, selon toute probabilité, provient la plus grande partie de notre matériel. L'indication de la présence de tourbe dans la vase pourrait s'interpréter comme correspondant au début de la remontée des eaux dans la vaste vallée que formait alors l'estuaire, processus qui s'est poursuivi par l'arrivée des eaux marines.

Enfin, 3 à 4 m de vase noire molle recouvrent l'ensemble précédent, pour former le fond de la baie d'Egoume.

3. LA MACROFAUNE DU PORT À BOIS

La liste systématique des Mollusques récoltés est donnée dans le tableau 1. Elle comprend 14 Gastropodes, 1 Scaphopode, 13 Lamellibranches, soit un total de 28 espèces.

La faune corallienne se réduit en fait à une seule espèce, *Schizoculina fissipara* M.E. & H.)[*], Madréporaire ahermatypique dont un échantillon (dont nous ignorons malheureusement la localisation précise) a été daté de 6430 ± 130 ans BP.[**]

L'étude écologique de la malacofaune révèle l'existence de 3 stocks essentiels, d'inégale importance: (1) un stock marin épilithe, peu diversifié spécifiquement; (2) un stock de tendance lagunaire (stock épimarin); (3) un stock marin psammophile.

3.1. *Le stock marin épilithe*

Cet ensemble, à vrai dire assez pauvre, comprend le Muricidae *Ocenebra inermicosta,* ordinairement lié aux rochers infralittoraux, et surtout le Bivalve *Pseudochama gryphina,* représenté par de nombreux individus.

Pseudochama gryphina est une espèce de très vaste répartition (Méditerrannée et Atlantique oriental, du Maroc à l'Afrique du Sud), dont la signification écologique est bien connue. C'est ainsi qu'en Méditerranée elle apparaît comme un des éléments de la 'biocoenose à Algues photophiles' (sensu Pérès & Picard 1964), communément rencontrée sur les surfaces rocheuses plus ou moins frotement éclairées, souvent associée à des Patelles et à des Vermets. Nous avons pu l'observer, au Sénégal (faciès rocheux de la presqu'île du Cap Vert), dans des conditions analogues. Sa distribution bathymétrique s'étend

---

[*]Détermination J.P.Chevalier (Muséum national d'Histoire Naturelle, Paris).
[**]J. & Y.Thommeret, Centre scientifique de Monaco (réf. MC-2377).

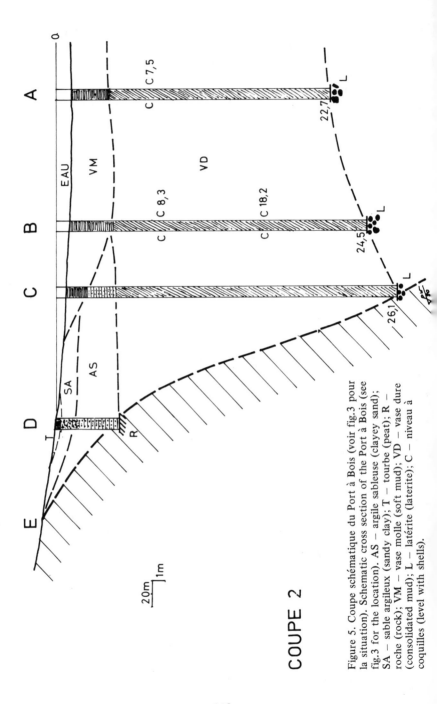

Figure 5. Coupe schématique du Port à Bois (voir fig.3 pour la situation). Schematic cross section of the Port à Bois (see fig.3 for the location). AS – argile sableuse (clayey sand); SA – sable argileux (sandy clay); T – tourbe (peat); R – roche (rock); VM – vase molle (soft mud); VD – vase dure (consolidated mud); L – latérite (laterite); C – niveau à coquilles (level with shells).

COUPE 2

de 1 à 50 m environ (nonobstant, bien entendu, de possibles variations locales liées à la transparence des eaux), ce qui correspond aux limites moyennes de l'étage infralittoral. Si l'on en juge par les documents publiés, *P.gryphina* a été draguée, en Afrique de l'Ouest, par l'expédition danoise 'Atlantide' jusqu'à une profondeur de 45 m (Nicklès 1955: 167), et par Albert de Monaco, à Santa Luzia du Cap Vert, jusqu'à 52 m (Dautzenberg 1910: 129). Toutes ces données sont parfaitement concordantes.

Bien que l'espèce soit typiquement inféodée aux substrats rigides offerts par les milieux rocheux, elle peut s'accommoder exceptionnellement d'un faciès de substitution. Certains peuplements de *Pseudochama* s'installent en effet sur des supports occasionnels: constructions portuaires, blocs immergés, mais aussi accumulations locales de coquilles mortes – à la condition toutefois que celles-ci bénéficient d'un amarrage suffisant et ne soient pas constamment déplacées par les mouvements de la nappe liquide. Il est vrai qu'en retour les *Pseudochama* peuvent contribuer, dans une certaine mesure, à assurer la stabilité du milieu, car leurs coquilles lourdes, fortement adhérentes au substrat, tendent à s'agréger entre elles au point de former quelquefois de véritables colonies de type pseudo-récifal. Par conséquent, la découverte de *Pseudochama* fossiles n'indique pas nécessairement la proximité d'un rivage rocheux, au sens strict du terme.

Il convient enfin de rattacher à ce stock marin épilithe *Nerita senegalensis*, qui appartient néanmoins à un autre horizon bionomique. En effet, ce petit prosobranche grégaire est, comme la majorité des *Nerita*, extrêmement commun à la limite supérieure des marées, et on peut y voir une espèce caractéristique exclusive des faciès rocheux médiolittoraux. Remarquablement résistant à l'exondation, il est capable de survivre à des émersions de 6 à 13 jours (Sourie 1954: 62). Il n'est pas impossible que les quelques exemplaires recueillis indiquent un bas niveau marin (mais encore faudrait-il disposer d'informations altimétriques et radiochronologiques précises) – à moins qu'ils n'aient été arrachés des niveaux médiolittoraux élevés, où ils ont dû vivre normalement, et entraînés par les courants.

3.2. *Le stock épimarin*

Cet ensemble, relativement important (en nombre d'espèces comme en nombre d'individus), comprend les formes habituelles des milieux margino-littoraux plus ou moins dessalés (lagunes côtières, estuaires, zones deltaïques). Les plus caractéristiques d'entre elles sont *Anadara senilis, Dosinia isocardia, Tympanotonus fuscatus*, auxquelles s'associent *Natica marochiensis, Thais forbesi, Thais callifera, Pugilina morio, Bulla striata adansoni, Felania diaphana*. L'Huître des Palétuviers, *Crassostrea gasar*, est aussi l'un des éléments les plus typiques du stock épimarin: on sait que ce Bivalve, dont l'euryhalinité est presque aussi étendue que celle d'*Anadara senilis*, trouve son optimum dans les mangroves, où il vit fixé en lourdes grappes sur les racines aériennes des Rhizophora.

119

En fait, quelle que soit la nature du lien qui les unit (rapports de dependance alimentaire, exigences édaphiques similaires), toutes ces espèces, quoiqu'euryhalines à différents degrés, forment un ensemble écologiquement très homogène. Elles entrent dans la composition de biocoenoses incontestablement apparentées, occupant dans l'espace des zones contiguës, et souvent réversibles (Elouard & Rosso 1977). A l'état fossile, elles sont abondamment représentées dans les faluns littoraux holocènes (Nouakchottien) de la Mauritanie et du Sénégal septentrional (Rosso *et al.* 1977). Le même type d'association a été retrouvé, plus récemment, au Sud-Bénin (Paradis 1978).

En ce qui concerne la signification bathymétrique de ce stock épimarin, il serait hasardeux d'avancer des chiffres précis. Nous pouvons dire toutefois que, si l'on écarte le cas de *Bulla striata adansoni*, relativement eurybathe, tous les Mollusques énumérés se cantonnent, à l'état vivant, dans les niveaux superficiels: étage médiolittoral, horizon supérieur de l'étage infralittoral. *Felania diaphana* et *Anadara senilis* ont été parfois rencontrées sur des fonds de 4 à 5 m, mais leur optimum bathymétrique se situe, en réalité, bien plus haut. D'autre part, *Crassostrea gasar* est typiquement médiolittorale, puisqu'elle vit normalement à la base des Palétuviers. De même, *Tympanotonus fuscatus* peut occuper des zones découvrant totalement à marée basse (Elouard & Rosso 1977). Enfin, *Pugilina morio, Natica marochiensis, Thais callifera, Thais forbesi* se trouvent habituellement en pleine eau, mais résistent sans dommage, enfouis dans l'épaisseur du substrat, à une émersion temporaire. Par conséquent, il semble que l'on soit en droit d'inférer, compte-tenu des possibilités extrêmes (0-5 m), une profondeur moyenne d'1 à 2 m.

3.3. *Le stock marin psammophile*

Ce troisième et dernier ensemble est riche, également, d'une dizaine d'espèces. Il comprend essentiellement les Gastropodes *Turritella ungulina, Archimediella annulata, A.gemmata, Genota mitraeformis*, et les Bivalves *Nuculana bicuspidata, Aequipecten flabellum, Clinocardium kobelti, Tellina senegambiensis, Corbula dautzenbergi*. Or ce stock franchement marin implique des conditions hydrologiques et bathymétriques bien différentes de celles qui ont dû régner lors du développement du stock précédent.

Toutes ces espèces, en effet, sont strictement sténohalines, sabulicoles (endopsammiques) et typiquement infralittorales. Leur présence exprime, à notre sens, la prédominance des influences marines sur les influences fluviales. Sans doute sont-elles contemporaines d'une partie au moins des malacocoenoses subfossiles de Port-Gentil (Nicklès 1952) et de Libreville (Rosso & Weydert 1979). Elles témoigneraient alors du même stationnement marin. Nous espérons pouvoir, par des mesures radiométriques ultérieures, leur assigner une place précise dans la chronologie.

Quant au Madréporaire *Schizoculina fissipara*, également connu le long des côtes américaines (Chevalier 1966), il paraît avoir des exigences proches de celles qui conditionnent la répartition de *Pseudochama gryphina*. Toutefois, sa localisation géographique plus étroitement limitée aux zones tropicales (golfe de Guinée, golfe du Mexique) révèle que l'espèce est plus rigoureusement sténotherme. Dans la région gabonaise, des *Schizoculina* sont communément rencontrés sur les fonds rocheux infralittoraux du cap Estérias, sur les blocs immergés (latérite) du Port à Bois, et même au débouché des grands chenaux de mangrove (baie de Mondah, par exemple), où la turbidité des eaux n'est pas trop élevée.

4. PLACE DE LA SÉRIE D'OWENDO DANS LE CYCLE HOLOCÈNE

Il est sans doute prématuré d'assigner à la série d'Owendo une place précise dans le cycle holocène. Le caractère indirect des observations, la rareté des datations, l'absence de critères altimétriques n'autorisent que des hypothèses soumises à un contrôle ultérieur. Il reste cependant loisible de reconstituer les grands traits de l'histoire récente à partir des informations acquises dans des secteurs proches, en particulier sur le littoral du Congo, du Cabinda, du Zaïre et de l'Angola (Giresse *et al.* 1974, Giresse 1975, 1978).

On admet (Giresse 1975) que le golfe de Guinée correspond à une zone déprimée, ainsi que le donnent à croire les mesures gravimétriques corrélées avec l'évaluation par satellite, qui compare le profil actuel à celui de l'ellipsoïde de référence (Gaposkin in Faure 1975). Le centre du golfe est, de ce fait, affecté par des anomalies d'altitudes négatives, à l'inverse de ce qu'on observe dans les secteurs Nord (Sénégal, Mauritanie) et Sud (Angola), où les dépôts littoraux quaternaires sont surélevés.

Ceci rappelé, et compte-tenu de l'homogénéité relative des conditions qui ont régné au sein du golfe (Giresse 1975), il n'est peut-être pas illégitime d'appliquer à l'estuaire du Gabon la chronologie établie pour l'estuaire du Kouilou (Congo). On obtiendrait alors la séquence suivante:

1. Le creusement de la vallée du Komo et les profils extrêmement abrupts des versants autour du Port à Bois seraient contemporains de la régression préholocène (dite aussi 'ogolienne', par référence à l'échelle stratigraphique admise en Mauritanie), datée de 18 000 ans BP. La mer est alors descendue aux alentours de -120 m. Les fleuves, très actifs, érodaient puissamment les continents, alors couverts de savanes. Il est à présumer que le remplissage latéritique des fonds de thalwegs (sur lesquels se sont arrêtés les sondages) s'est opéré au cours de cette période, où le climat était relativement sec, avec cependant des saisons contrastées (ce qui n'exclut pas des crues parfois violentes).

2. La transgression holocène (ou 'nouakchottienne' selon la terminologie stratigraphique adoptée en Mauritanie), entre 8 000 et 5 000 ans BP environ, correspond à une période de réchauffement qui va de pair avec une augmentation de l'humidification. Les eaux transgressives, chaudes, favorisent l'implantation d'une malacofaune spécifiquement diversifiée. De cette période datent

121

probablement les Mollusques sténohalins de la Lowé (Rosso & Weydert 1979) et d'Owendo (stock marin psammophile), ainsi que les colonies coralliennes à *Schizoculina fissipara*.

3. La fin de cette phase transgressive est marquée par une augmentation de la pluviosité. La forêt remplace la savane; les mangroves se développent largement, comme en témoigne le développement de *Crassostrea gasar*. Le type de sédimentation évolue alors progressivement vers les faciès vaseux, avec prédominance de conditions estuariennes ou lagunaires (dessalure, confinement relatif du milieu) favorables à l'installation de la malacocoenose typique à *Anadara senilis, Dosinia isocardia, Felania diaphana, Natica marochiensis* (Elouard & Rosso 1977), ce qui correspond précisément au 'stock épimarin' d'Owendo. La datation effectuée sur l'un des éléments de ce stock, *Thais callifera,* a donné un âge de 3 040 ± 70 ans BP.

Ces conditions sont à peu près celles qui prévalent encore aujourd'hui.

4. Les vases molles susjentes ont probablement été mises en place lors d'un dernier retour transgressif, entre 3 000 et 500 ans BP si l'on se réfère, une fois de plus, au cadre chronologique esquissé par Giresse (1975). Cependant, rien n'indique ici que ce mouvement transgressif a dépassé le précédent en amplitude. De même, aucun argument ne peut être invoqué pour inférer une légère régression ('taffolienne' dans la terminologie stratigraphique élaborée pour les dépôts mauritaniens) entre ces deux épisodes. On peut seulement constater la différence de compaction des remplissages.

CONCLUSION

Si la macrofaune subfossile du Port à Bois d'Owendo est spécifiquement moins riche que celle de la Lowé (60 espèces) et, a fortiori, de Port-Gentil (117 espèces, étudiées par Nicklès 1952), elle n'en apporte pas moins un complément appréciable à la connaissance systématique des thanatocoenoses quaternaires du Gabon. Il est encore trop tôt pour tenter des corrélations précises entre les divers sites prospectés. Néanmoins, l'examen des listes publiées révèle, ici comme là, un certain nombre de 'hiatus écologiques', solidaires de modifications intervenues dans les milieux, qui manifestent l'existence de plusieurs épisodes successifs, marquant les principales phases de la transgression holocène.

REFERENCES

Bloom, A.L. 1977. Atlas of sea-level curves (IGCP Project 61). Polycop.
Bouma, M., M.Lanau, Ch.Perrusset & P.Weydert 1978. Nouvelle esquisse géologique de la partie nord du bassin côtier gabonais grâce à l'imagerie radar. *CR Ac. Sci. Paris* 287(D): 215-218.
Brink, A.H. 1974. Petroleum geology of Gabon Basin. *Am. Ass. Petr. Geol. Bull.* 38(2): 216-243.

Carte marine no.6369, 1963. Estuaire du Gabon.

Carte marine no.6360, 1973. Cap Estérias à Iguéla.

Chevalier, J.P. 1966. Contribution à l'étude des Madréporaires des côtes occidentales de l'Afrique tropicale. *Bull. IFAN* A28(3): 912-975 (1º partie). *Bull. IFAN* A28(4): 1356-1405.

Dautzenberg, Ph. 1910. Contribution à la faune malacologique de l'Afrique occidentale. *Actes Soc. linn. Bordeaux* 54:1-174, 4pl.

Dwars 1967. *Le Port d'Owendo: rapport particulier no.2.* Annexe A: recherche géotechnique. Minist. Trav. publ. gabonais & FED.

Elouard, P. & J.C.Rosso 1977. Biogéographie et habitat des Mollusques laguno-marins du delta du Saloum (Sénégal). *Atti Soc. Ital. Sci. nat. Museo civ. Stor. nat. Milano* 118(2): 165-184.

Faure, H. 1975. New approach to quaternary sea-level problems. *Ass. Sénég. Et. Quatern. afr. Bull. Liaison* 44-45: 81-84.

Giresse, P. 1975. Nouveaux aspects concernant le Quaternaire littoral et sous-marin du secteur Gabon-Congo-Cabinda-Zaïre et accessoirement de l'Angola. *Ass. Sénég. et. Quatern. Afr. Bull. Liaison* 46: 45-52.

Giresse, P. 1978. Le contrôle climatique de la sédimentation marine et continentale en Afrique centrale atlantique à la fin du Quaternaire. Problèmes de corrélations. *Paleogeogr., Paleoclimatol., Paleoecolog.* 23: 57-77.

Giresse, P. & G.Kouyoumontzakis 1974. Observations sur le Quaternaire côtier et sous-marin du Congo et des régions limitrophes. Aspects eustatiques et climatiques. *Ass. Sénég. Et. Quatern. Afr. Bull. Liaison* 42-43: 45-61.

Hourcq, V. & Hausknecht 1954. *Notice explicative sur la feuille Libreville-ouest au 1/500 000.* Imprimerie Nationale, Paris.

Hudeley, H. 1970. Notice explicative de la carte géologique de la République Gabonaise au 1/1 000 000. *Mém. BRGM* 72.

Laboratoire central d'hydraulique de France 1977. *Extension du port d'Owendo, Rapport général* (1º partie). Bur. Centr. Et. Equip. Outre-mer et OPRAG.

Lehner, P. & P.A.C.de Ruiter 1977. Structural history of atlantic margin of Africa. *Am. Ass. Petr. Geol. Bull.* 61(7): 961-981.

Nicklès, M. 1952. Mollusques du Quaternaire marin de port-Gentil (Gabon). *Bull. Dir. Mines Géol. AEF* 5: 75-101.

Nicklès, M. 1955. Scaphopodes et Lamellibranches récoltés dans l'Ouest africain. *Atlantide Report* 3: 93-237. (Danish science press).

Paradis, G. 1978. Interprétation paléoécologique et paléogéographique des taphocénoses de l'Holocène récent du Sud-Bénin, à partir de la répartition actuelle des Mollusques littoraux et lagunaires d'Afrique occidentale. *Géobios* 11(6): 867-891.

Pérès, J.M. & J.Picard 1964. Nouveau manuel de bionomie benthique de la mer Méditerranée. *Rec. Trav. Stn mar. Endoume* 47(31): 1-137.

Rosso, J.C., P.Elouard & J.Monteillet 1977. Mollusques du Nouakchottien (Mauritanie et Sénégal septentrional): inventaire systématique et esquisse paléoécologique. *Bull. IFAN* A39(3): 465-486.

Rosso, J.C. & P.Weydert 1979. Thanatocoenoses marines quaternaires du Gabon: inventaire et signification écologique de la malacofaune. *Géobios* 12(1): 133-136.

Sourie, R. 1954. Contribution à l'étude écologique des côtes rocheuses du Sénégal. *Mém. IFAN*, Dakar 38: 1-342, 23 pl.

Steinfeld, K. 1972. *Le port d'Owendo: rapport géotechnique.* Erdbau Laboratorium, Hamburg.

Weydert, P. 1979. *Eléments de géologie gabonaise.* Publ. ronéo DEP Libreville, 30pp.

SPATIAL MANAGEMENT OF HOMINID GROUPS AT OLDUVAI: A PRELIMINARY EXERCISE

MILLA Y. OHEL

Department of Sociology & Anthropology, University of Haifa, Israel

ABSTRACT

Raw material exploitation for the manufacture of lithic artifacts by settlements of Beds I and II at Olduvai Gorge is initially explored applying some approaches from cultural geography. Overall and selective preferences by hominid groups with relation to quality of raw material and distance from resources are discussed and priorities in decision-making concerning the location of settlements are suggested. Owing to the time-span involved (2 to 1 million years ago) and the constraints of the data, merely tentative interpretations may be advanced. It would appear that proximity of fresh water and food resources weighed more in the decision-making process than that of raw materials.

INTRODUCTION

Environmental cognition and spatial organization in the sense in which they are presented within the field of human geography (e.g. Lynch 1960, 1972, Haggett 1966, Berry 1967, 1972, Gould & White 1974) are out of reach for prehistoric studies. Human geography is based on direct communication with living individuals, prehistoric analysis must rest on 'mute' and 'blind' data, obliging indirect inferences.

Nevertheless, some insights may be gained from human cognitive geographical enterprises, which in turn may stimulate new trajectories of exploration into palaeolithic times. The present article is one such attempt. This attempt is neither unique nor innovative in applying human geographic procedures to prehistory. However, apart from a few published studies (Stiles et al. 1974, Larson 1975, Hay 1976, Speth & Davis 1976) I am unaware of a similar attempt concerned with such an ancient period (2 to 1 million years ago). The present analysis is a preliminary exercise, since the constraints of the data must as yet deter too-confident inferences (Hodder & Orton 1976). Sophisticated statistics have deliberately not been pursued.

THE OLDUVAI SITES

Nineteen levels (or compounds of levels) from Beds I and II of Olduvai have been selected for the present study, as follows:

1. DK, Levels 1-3
2. FLK, 'Zinjanthropus' Floor
3. FLK N, Level 5
4. FLK N, Level 3
5. FLK N, Levels 1/2
6. HWK E, Level 1
7. HWK E, Level 2
8. HWK E, Level 3
9. HWK E, Level 4
10. FLK N, Sandy Conglomerate
11. MNK, Skull Site
12. EF-HR
13. FC W, Floor
14. FC W, Tuff
15. MNK, M.O.S. (= Main Occupation Site)
16. SHK, all levels
17. TK, Lower Floor
18. TK, Upper Floor
19. BK

BED	PART OF BED	MARKER TUFFS	DATING (m. y. a.)	SITE	LITHO-FACIES (EASTERN)
BED II (MIDDLE)	UPPER		1.15	BK, TK	FLUVIAL - LACUSTRINE
	MIDDLE	II D		SHK, MNK, M.O.S., FC W, EF-HR	FLUVIAL - LACUSTRINE
		II B	≤ 1.60	MNK, Skull Site, FLK N: Sandy Cong.	
		LEMUTA MEMBER, II A		HWK E: Level 4, HWK E: Level 3	
	LOWER		1.71	HWK E: Level 2, HWK E: Level 1	LAKE MARGIN
		I F	1.70		
BED I	UPPER	I C		FLK N: Levels 1-6	LAKE MARGIN
		I B	1.79	FLK: 'Zinj.' Floor	
			1.90	DK: Levels 1-3	

Figure 1. Schematic stratigraphic position of the sites studied, Olduvai, Beds I and II, and their distribution by lithofacies (modified after Leakey 1971: Table 1 and Hay 1976; not all tuffs are included).

126

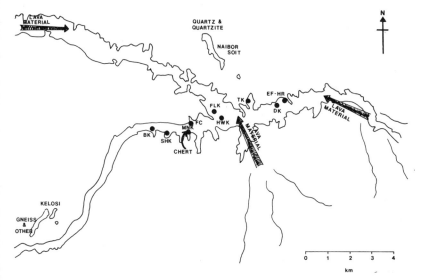

Figure 2. Sketch map of Olduvai Gorge showing location of sites discussed, location of raw material resources, and presumed directions of streams that brought lava material into the gorge (modified after map and Figure 4 in Leakey 1971).

These belong to nine site localities, the stratigraphic positions of which are shown on Figure 1. A sketch map showing the localities, resources of raw materials, and presumed directions of streams that brought lava into the gorge, is provided in Figure 2.

As seen on Figure 2, the rough distance between extremeties is 6-7 km. It is impossible to establish temporal synchronization between sites owing to the time spans involved (Figure 1). Consequently, considerations regarding spatial arrangements pertaining to the existence and activities of localities at one and the same time, must be neglected.

The sites were situated either in the margin zone of an ancient shallow lake or near ponds and pools in lacustrine environments of an alluvial fan, or on stream banks. Throughout the nearly one million years of Beds I and II, the lake level had been fluctuating to considerable extents (average diameter 7-25 km). It is suggested (Leakey 1971, Hay 1976) that the lake reached maximum level early in Bed I times prior to the penetration into the gorge of basalt flows. Later the lake shrank gradually to its minimum limits during Bed I (between Tuffs ID and IF times), subsequently rising rather abruptly anew at the end of Bed I (Tuff IF) but achieving only about half of its former largest dimensions. These conditions prevailed during the lower part of Bed II until the disconformity (1.60 m.y.a.), thereafter shrinking again in remaining Bed II times to disappear entirely shortly before the end of Bed II deposition.

127

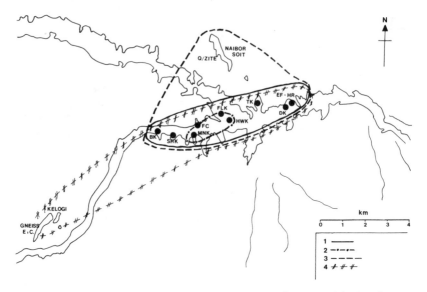

Figure 3. 'Fields' encompassing the sites and the resources of raw materials: 1 – the nearest, of lava; 2 – the nearest, of chert (from MNK used during a restricted time at HWK and FLK); 3 – the second-nearest, of q/zite; 4 – the remote, of gneiss, etc.

Thus it is likely, for instance, that DK and EF-HR, while situated close to each other geographically (Figure 2), experienced different environmental circumstances. DK, at the time of occupation, was located on an actual lake shore. EF-HR, on the other hand, while approximately upon the same spot about half a million years later, was at some distance from the lake shore of its time.

RAW MATERIALS

The principal rocks utilized were lava, quartz and quartzite, gneiss, and chert.

The main variety of lava used was Sadiman lava (Hay 1976). This is a dense, homogeneous, fine-grained rock, green coloured, that was abundantly available in the form of cobbles in stream channels within a distance of 2 km from all sites (Figure 2). Quartz and coarse quartzite (mostly of the green and brown varieties) were attained from Naibor Soit (Figure 2) at an approximate distance of 2 to 5 km. Gneiss, and a few other rare rocks, were imported from Kelogi (Figure 2) at an approximate distance of 6 to 13 km. Chert, when it became exposed at MNK, occurred mostly as nodules, white and opaque or milky and translucent. This fine material which 'consists of microcrystalline quartz and resembles the flints of England and the mid-continental USA in mechanical properties' (Hay 1976:184) was exploited extensively by hominids from the HWK and FLK sites at a distance of 1 to 1.5 km.

The degree of natural disturbances in walking from any one site to the resources of raw material and in carrying back stones, blanks, or artifacts, cannot be evaluated for the time being. Such disturbances which might have played a role with relation to 'distance friction' (or 'payoff') must therefore be disregarded. Yet it seems feasible to draw schematically the diverse 'fields' as related to the sites as a whole unit, and the raw material resources. Figure 3 indicates that (a) the lava 'field' is the nearest; (b) the chert 'field' is also the nearest at a restricted period; (c) the quartz/quartzite (q/zite) 'field' is the second-nearest; and (d) the gneiss, etc. 'field' is the remote one.

Mechanical properties (physical quality) of the rock types with regard to workmanship are not easy to assess. Although a universal grading may be suggested, such a grading suffers from undue generalizations, since it can hardly account for local diversities. Thus, although universal grading was here considered, primary attention was paid to statements or remarks in the site reports (Leakey 1971, Hay 1976, also, the grading that follows was generally accepted by Hay in a personal letter dated 27 August 1975).

Fortunately, the incidence of chert occurring during a limited period in the lower part of Bed II has indicated a satisfactory solution. This chert was of far better quality than all other rocks, attested to by the extensive utilisation of chert while available, as will be shown. The physical qualities were subsequently graded as follows: Chert: high quality (HQ); Lava: medium quality (MQ); Q/zite: low quality (LQ); Gneiss, etc.: low quality (LQ).

MODEL AND ASSUMPTIONS

Drawing from a heuristic paradigm was presented by Berry (1972), a heuristic flow model is here outlined (Figure 4). While Berry's model attempts to encompass a complete ecosystem, this is a partial model, the lack of reliable information obliging the theoretical scheme to postulate merely one facet of an ecosystem as related to a few other facets.

In the framework of the proposed model, several assumptions may be outlined:

1. The lithic assemblage from each level represents a functional relationship between the artifacts contained and the hominid group(s) which produced them.

2. The preferences in the utilization of different raw materials are reflected by the diverse proportions of artifact categories as represented by each level (= group = settlement). These preferences interact with the distances of the rock resources and the physical quality of the rocks, as follows.

3. The nearest resource will regularly be exploited to a greater extent than the remoter resources. This situation will prevail even when the rock provided

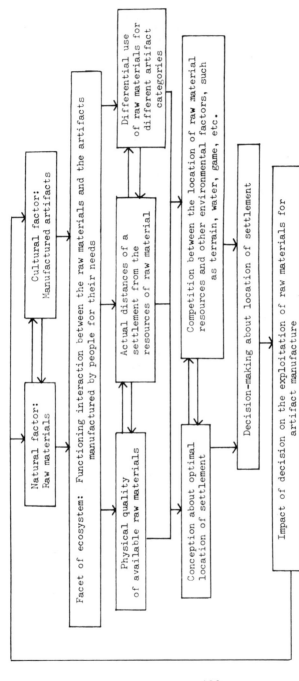

Figure 4. Heuristic flow model representing assumed interaction within a facet of an ecosystem.

by the nearest resource is not of the highest quality available within the broader vicinity (cf. Hodder 1977). By the same token the next-nearest resource will be utilized to the next-greatest extent, etc.; and the remotest resource to the smallest extent. This assumption is based on the 'law of minimum effort' (Haggett 1966) which seems to coincide well with hunter-gatherer groups (Jochim 1976). Assuming the absence of containers and any means of accessory transportation, it seems reasonable to suppose that, ordinarily, distance would overrule quality.

4. Resources of suitable raw material apparently strongly influenced decisions about settlement placement. However, even in such a partial ecosystemic scheme as outlined here, several other considerations must be accounted for (Hodder & Orton 1976): appropriate terrain to stay, rest and move upon; a protectable and topographically advantageous spot; proximity of drinkable water; availability of game and edible plants – such might have well been competing with rock resources in the decision-making process as to what precise location to choose (Jochim 1976, with references).

The assumption here is that ready accessibility to water would have had

Table 1. Percentages of different raw materials exploited for artifact manufacture by settlements at Olduvai, Beds I and II.

Settlement	Lava (MQ)	Q/zite (LQ)	Chert (HQ)	Gneiss, etc. (LQ)
1. DK, Levels 1-3	72.5	27.5	0.0[a]	–
2. FLK, Zinj Floor	26.6	72.9	–	0.5
3. FLK N, Level 5	60.7	39.3	–	–
4. FLK N, Level 3	44.3	55.7	–	–
5. FLK N, Levels 1/2	37.2	62.6	–	0.2
6. HWK E, Level 1	63.9	35.8	–	0.7
7. HWK E, Level 2	30.8	51.4	17.8	–
8. HWK E, Level 3	26.8	38.7	33.6	0.9
9. HWK E, Level 4 (Sandy Conglomerate)	23.9	39.2	35.6	1.3[b]
10. FLK N, Sandy conglomerate	24.5	32.3	43.2	–
11. MNK, Skull Site	43.5	55.4	–	1.1
12. EF-HR	67.6	30.0	2.2[c]	0.2[d]
13. FC W, Floor	25.0	74.3	0.0[e]	0.7[f]
14. FC W, Tuff	22.7	76.2	–	1.1[f]
15. MNK, M.O.S.	22.1	76.9	–	1.0[f]
16. SHK, all levels	9.5	90.0	0.2	0.3[g]
17. TK, Lower Floor	8.1	91.4	–	0.5
18. TK, Upper Floor	15.6	84.2	–	0.2
19. BK	17.5	80.3	0.1	2.1[g]

a – one rolled flake
b – migmatite
c – five pieces from surface, source unknown
d – pegmatite

e – one waste piece
f – includes pegmatite
g – includes pegmatite and welded tuff

the strongest influence upon decision. The decision would have in turn exercised its impact on the pattern of raw material exploitation. Visualizing the Olduvai groups as of 'primitive', relatively modest requirements (low 'aspiration level': Jochim 1976) and of a 'primitive', opportunistic technology (Freeman 1975, Ohel 1977), it is assumed that they were sufficiently satisfied with the raw materials closest at hand.

DATA ANALYSIS

Table 1 provides a breakdown of the raw materials utilized by each settlement (= level). Such data are not given in a complete, summarized form by Leakey

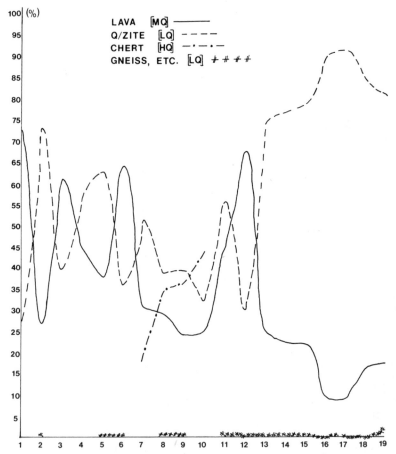

Figure 5. Exploitation of different raw materials for artifact manufacture by the various settlements (for identification of settlement numbers, see Table 1).

132

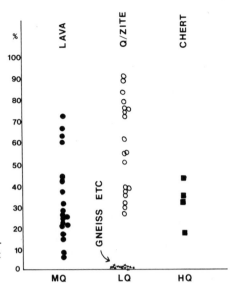

Figure 6. Preferences in exploiting different raw materials according to their quality for manufacture of artifacts (sign represent settlements).

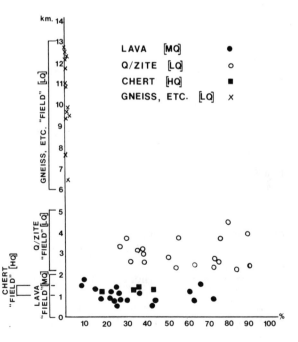

Figure 7. Distribution of raw materials used for artifact manufacture plotted against distances of settlements from the resources of raw material.

133

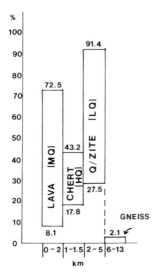

Figure 8. Histograms showing ranges of exploitation of raw materials against ranges of distances of the settlements from the resources.

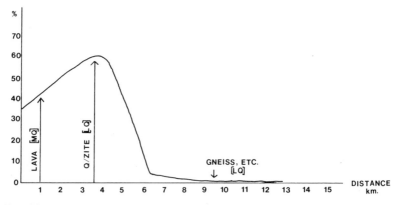

Figure 9. Trend of exploitation of raw materials by all settlements as a whole (percentages represent means) plotted against distance from resources — excluding chert.

(1971) and had to be calculated from counts throughout the report and in particular from her Table 7. Table 1 was then transformed into a curve plot (Figure 5) to facilitate easier appreciation of the way in which the raw materials were exploited throughout all settlements through time (for time spans, see Figure 1).

While a fair balance (counter-dovetailing) is demonstrated on both extremeties of Figure 5 between lava and q/zite, this is distorted in the middle by the

134

intrusion of chert (7-10). A few additional phenomena are worth noting: (a) lava achieves no more than four peaks throughout (1, 3, 6 and 12); (b) a decline is marked in lava when chert intervenes compared to some fluctuations on the q/zite's part (7-10); and (c) a pronounced dominance of q/zite over lava at the righthand side (13-19) is manifested. These phenomena will be attended to later.

Figure 6 illustrates the preferences in the exploitations of the raw materials with regard to physical quality. Chert, as mentioned, was exposed during a limited time. Nonetheless, even then, and in spite of it being of the best quality available, chert never attained as high proportions as did lava or q/zite at some other periods. Between these latter two an interesting relationship is suggested: namely, that although q/zite is judged of a lesser quality than lava, yet it was generally utilized to a greater extent. The explanation may perhaps rest in the preference of q/zite for special categories of artifacts (Leakey 1971, Hay 1976) – a point to which we will return.

Figure 7 shows the distribution of raw materials with relation to distances of settlements from resources. Most of the lava signs are packed while the q/zite ones, although reaching in many cases higher proportions, are more dispersed. This appears to be explainable by the fact that lava was obtained in the immediate vicinity of every settlement, whereas q/zite was not. Q/zite was attained from farther away (Naibor, see Figure 3) and, therefore, stands in disparate distance relationships with each one of the settlements. The state of the gneiss signs is clearcut: both low quality and great distance (Kelogi, see Figure 3) induced almost negligible exploitation. Figure 8 presents the data contained in the previous figure more succinctly.

Figure 9 shows the trend of exploitation with relation to distance of the settlements as a whole unit from the resources, chert excluded. Figure 10 demonstrates a similar curve just for those four settlements (as a whole unit)

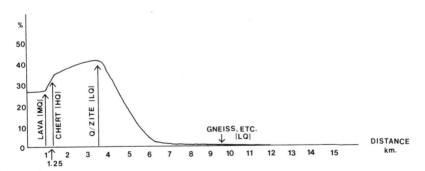

Figure 10. Trend of exploitation of raw materials by the four settlements utilizing chert during the restricted period it became available, plotted against distance from resources (percentages represent means).

for which chert was accessible. The percentages are calculated to approximate means.

From Figure 9 it seems apparent that although lava is of a better quality than q/zite, and is the nearer resource, it was q/zite that was utilized more. When the high quality chert became procurable, and within the immediate vicinity (1-1.5 km) of the contemporaneous settlements (7-10, see Table 1), it supplanted lava in particular, and to some extent q/zite as well (Figure 10). Yet, the trend of the curve reflecting that limited period does not exhibit drastic departure from the curve in Figure 9.

Figures 9 and 10 may be envisaged as representing a distance-decay process (e.g. Haggett 1966, Wilson 1971), except that q/zite, which is the second-nearest resource, achieves the peak in the curves, thus introducing some 'abnormality' compared to the commonly accepted pattern of normal distance decay. This abnormality also refutes our assumption that the nearest resource would be utilized to the greatest degree, and that distance would overrule quality in detering exploitation. We still lack some essential data for suggesting a sound explanation of the above discrepancy. The following considerations ought, however, to be born in mind with connection to the distance-decay curves.

1. The nomination of Naibor (q/zite) as the second-nearest resource may have been wrong. For some of the settlements Naibor might have been as close as lava inside the gorge, if not closer (Figure 3, also Hay 1976:184). Having been washed into the gorge from all directions (Figure 2), lava cobbles and pebbles were assumed to have been constantly at hand, an assumption that may not hold for all settlements. Thus the generalized assumption could have generated the 'abnormality'.

2. The distance to Naibor ranges from 2 to 5 km. We still do not know whether such a range was appraised by the hominids as a 'greater distance' than the resources in the gorge. Recent studies about 'least effort' and 'lapse rates' (ethnographic hunter-gatherers' included) might not have applied constantly and regularly to those creatures 1-2 m.y.a., at least when relatively short distances were involved (and see Reidhead 1979, with references).

3. The assumptions may be mainly correct; yet for reasons unknown, the Olduvai groups preferred q/zite to lava even though the former was of worse quality and more remote than the latter.

4. Finally, an error in 'sampling' is feasible: the high proportions of q/zite artifacts may be due to their being mostly the smaller artifacts, and therefore of far greater numbers than the lava ones (Figure 12, and Leakey 1971). Hominids might have had no need to go too frequently to Naibor, for whenever they did go and fetch q/zite pieces, these were quite sufficient to yield a greater number of small artifacts with comparison to larger tools made from the immediately available lava.

In sum, although the assumptions cannot be either confirmed or rejected categorically, some of the problems to be faced in attempting a satisfactory solution have been demonstrated.

OVERALL PREFERENCES

The construction of mental maps (Gould & White 1974) cannot be fully duplicated here. Yet, their basic idea and the primary building blocks may be adopted.

The first step is to rank a series of areas in accordance with individual preferences. The second step is to convert these preferences into a matrix of correlation coefficients (for the procedure, see, e.g. Thomas 1976). Taken as unit vectors (third step), the coefficients can be illustrated as cosines of angles between vectors representing the individuals involved in the preference rankings. The ensuing steps, aimed at collapsing all individual preferences into one mental map, are impractical here. Dealing with three to four raw material resources (= areas) as ranked by settlements/groups (= individuals), most of which make their choice of either one or two materials (lava or q/zite), does not require the collapsing step. The overall preferences and their interrelationships ought to be clear from the vectors alone.

However, one preliminary step seems inescapable. That is, all 19 groups cannot be intercorrelated as a whole because of the intervention of chert (in 7-10), introducing a resource, the information of which is not shared by most groups. Consequently, the 19 are divided into three groupings: I: 1-6; II: 7-10; III: 11-19 (Figure 11). This division appears to be justified also from Figure 5. Note that gneiss, etc., even when not represented at all in a level (Table 1) is assigned the lowest rank on the assumption that the resource was known to these groups as well but not exploited.

Figure 11 assists us in recognizing the basic tendencies of preferences within each of the three groupings. Further, by displaying the standing of each settlement/group in relation to the others within each grouping, the vectors enable an appraisal of the feedback between the factors discussed earlier, such as specific geographical location, resources of materials, and their distances from settlements.

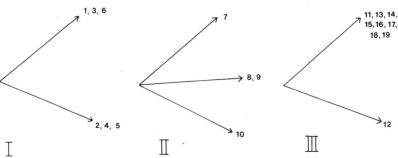

Figure 11. Correlation coefficients shown as cosines of angles: major groupings of Olduvai settlements representing interrelationships of preference of raw material resources.

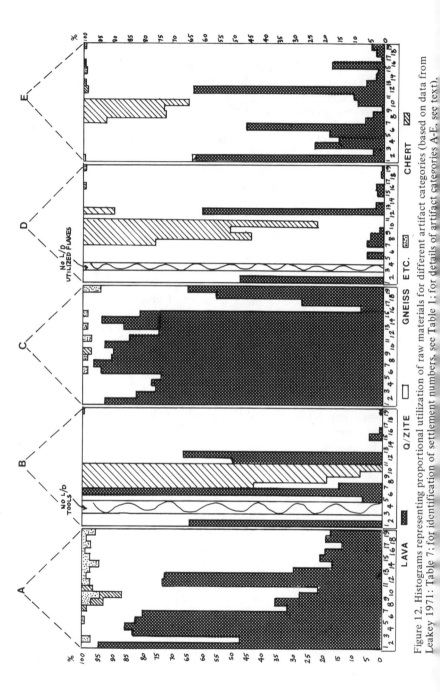

Figure 12. Histograms representing proportional utilization of raw materials for different artifact categories (based on data from Leakey 1971: Table 7; for identification of settlement numbers, see Table 1; for details of artifact categories A-E, see text).

Three settlements of grouping I (1, 3 and 6) prefer lava while another three (2, 4 and 5) prefer q/zite. In Figures 1-3, no dichotomy can be discerned between the two subgroupings with respect to temporal or spatial layouts. All six settlements interfinger along the time span of the sequence (Figure 1). Any marked differences in distance either to lava or q/zite resources can hardly be invoked (Figure 3), all six belonging to sites within close proximity (Figure 2). Yet, three prefer lava while the remaining three q/zite (Figure 11). The explanation must therefore be sought apart from distance to resources.

In grouping II a different pattern is apparent. Settlements 7, 8, 9 and 10 belong to two sites situated closely to each other and of nearly equidistance from all resources, including chert at MNK (Figure 3). However, Table 1 and Figure 11 show a gradual shift in preferences through time (Figure1): from q/zite still preferred by 7, through almost an equilibrium between q/zite and chert later in 8 and 9, finally to a preponderance of chert in the last settlement(10). Again, distance cannot fully explain this alteration in preference of the four settlements.

Grouping III reveals a divergent trend of preferences from the former two. Apart from settlement 12 that prefers lava, all other eight: 11, 13-19 prefer q/zite (Table 1, Figures 5 and 11). These are spread throughout the upper half of Middle Bed II and Upper Bed II (Figure 1) as well as over the entire area (Figure 3). Distance can hardly be called upon as a decisive factor in the preferences of grouping III as well.

In sum, we are led to further loss of ground to the assumption of a necessary negative correlation between distance of resource and extent of its exploitation. In other words, it is ostensibly not always the rule that the nearest resource will be exploited most. This observation applies to the close vicinity alone, say up to 5-7 km radius (see on the 'plateau effect' in Plog 1976:258, with references).

SELECTIVE PREFERENCES

Figure 12 represents the proportions of raw materials as utilized for different artifacts categories by settlements, based on data from Leakey (1971:Table 7). The categories follow her division:

A: choppers, 'proto-bifaces', polyhedrons, discoids, spheroids, sub-spheroids, and heavy-duty scrapers.

B: light-duty scrapers, burins, awls, laterally trimmed flakes, sundry small tools.

C: anvils, hammerstones, utilized cobbles, nodules, and blocks.

D: light-duty utilized flakes, etc.

E: debitage.

Space limitations prevent me from discussing here in detail each one of the settlements. Instead, only a brief concluding summary for groupings will be provided.

139

Grouping I. (a) In general, lava was preferred for large, heavy-duty artifacts while q/zite for small, light-duty ones; (b) in a few cases, however, deviating selective preferences are manifested, which even result sometimes in dichotomous overall preferences (Figure 11); (c) these manifestations introduce an incremental factor, namely the preferential selectivity of raw materials in line with the specific differential functions as assigned to diversified artifact categories. This latter fact might have born far more weight in the selection process of the available materials than either physical quality of rock or distance to resources, provided this distance did not exceed reasonable proximity (say 5 to 7 km). Although the specific functions cannot be ascertained, their existence is reflective by the selective preferences.

Grouping II. (a) With chert obtainable, it was maximally exploited for small, light-duty artifacts; (b) q/zite mostly replaced lava for large, heavy-duty tools; (c) if indeed q/zite was remoter than lava, the groups — for one reason or another — did not consider this distance a hindrance.

Grouping III. (a) A great deal of similarity is displayed: low proportions of lava are involved in the production of large, heavy-duty tools, these being made mostly of q/zite, while small, light-duty artifacts are almost entirely made of q/zite; (b) again, even if q/zite was a farther resource than lava, the former was preferred by most settlements.

Thus, the preliminary assumptions are not unequivocally supported by the data. In fact, three situations ought to be distinguished. First, when the lowest-quality material (gneiss, etc.) was at relatively great distance, it was least exploited. Second, when the highest-quality material (chert) was in the immediate vicinity, it was exploited maximally, but merely for certain artifact categories. Third, when a medium-quality material (lava) and a low-quality one (q/zite) were at relatively short distances, the slight difference in distance, whenever existing, did not play a decisive role in determining exploitation. What played the decisive role in this latter situation was the selective preference for one material or the other with relation to one artifact category or another as differentially found appropriate by the various settlements.

The opportunity to make the first situation to conform to a distance decay curve proper cannot be tested because we lack far-fetched high-quality products in any convincing quantities. As for the two remaining situations, if the q/zite resource is taken to be second-nearest, then the Olduvai distance decay curve does not match precisely the ideal or common one (and see again Figures 9 and 10).

COMPETING CONSIDERATIONS

Raw material resources at Olduvai do not appear to have competed strongly against some other crucial considerations in the decision-making process leading to the location of settlements, as long as some sorts of suitable raw materials were obtainable within a reasonably near vicinity. The comparatively low-ranking of such resources on the scale of vital survival requirements may be even more appreciated when the importance of the other requirements is duly stressed.

Figure 1 shows that all settlements without exception were located within the close neighbourhood of water. Further, even though the Olduvai lake was a saline, alkaline lake, its southeastern margins were subjected to relatively fresh water floodings, and it is the eastern zone of the basin and along the southern margin of the lake where all settlements were situated (Figure 1; Hay 1976:3, 25 and Appendix B, Leakey 1975:479, Butzer 1977:576). Thus it is clear that one of the most critical, if not the critical condition for settlement location was the proximity of fresh water, especially in a macro-environment of arid or semiarid climate (Hay 1976:22).

A major change in the setting of the basin occurred with the disconformity (about 1.60 m.y.a.). Prior to the disconformity all archaeological sites are found within the lake margin facies. With all its fluctuations during this time span, the lake continued to be the major, if not the exclusive focus, of hominid occupation. All settlements of groupings I and II (1-10) belong in this phase. The disconformity marked the onset of widespread faulting and folding in the central part of the basin; the size of the lake decreased drastically and an increase of grassland ensued (Leakey 1971:260, Hay 1976:26). Most hominid groups, including all grouping III members (11-19), now occupied the fluvial-lacustrine facies.

This major break apparently also caused a great biotic change which exercised an important influence on the pattern of hominid subsistence. Prior to the disconformity the faunal assemblages were dominated by water- and swamp-dwelling species, while after it the proportion of plains animals favoring open savanna and riverine conditions increased abruptly and dominated the assemblages (Leakey 1971:*passim*, Hay 1976:*passim*). Hominid groups, then, chose those locations from which they could best control the movements of animals and easily obtain their staples. The competitive power of food approachability weighted more than that of raw material in the decision-making process toward settlement location.

It seems that the ranking of priorities regarding settlement location took the following order: A. Fresh water; B. Food staples; C. Raw materials.

How to relate this conception to the selective preferences of raw materials discussed above remains a tough question. Leakey (1971, 1975, 1976) envisages four idiosyncratic technologies along Beds I and II: Oldowan, Developed Oldowan A and B, and Early Acheulean. The Oldowan is confined to Upper Bed I and Lower Bed II, the Acheulean to Middle and Upper Bed II while the Developed Oldowan runs parallel to the Acheulean. Leakey (1971:269) seems to believe that the Acheulean is an intrusive element.

However, Leakey's two Developed Oldowan technologies are not completely unquestionable; a critical argumentation in this respect will require

141

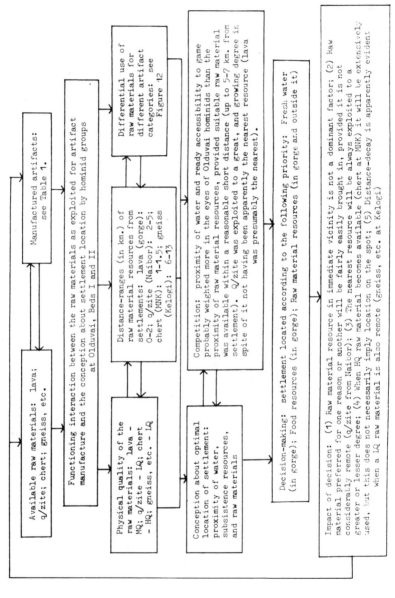

Figure 13. Interpreted scheme representing a suggested interaction flow chart of a facet of an ecosystem at Olduvai, Beds I and II (cf. Figure 4).

142

ample space and will not be pursued here (see Stiles 1979 where her classification is refuted). Some clue may possibly be detected through a different approach applied in a previous study of mine (Ohel 1977). There I examined the association of two categories of bone (small bones and bone fragments versus larger bones) and two major categories of artifacts (debitage versus other tools) within two different shapes of concentration (circular versus elongated) as observed upon 14 living floors in Beds I and II. I very cautiously suggested there to regard the elongated concentrations characterized by large bones associated with a few large tools as initial food-processing areas, and the circular concentrations characterized by bone fragments associated with small artifacts as final food-consuming areas.

What may be of interest to the present discussion is the observation I made in a draft version of that study. While on Oldowan floors (which for me include also Leakey's Developed Oldowan A occurrences) there were discernible both circular and elongated concentrations, on Early Acheulean floors (which for me include also Leakey's Developed Oldowan B occurrences) elongated concentrations were completely absent, circular ones alone present. My tentative conclusion was that whereas the Oldowans used to dismember the animals close-at-hand to the areas of consumption, the Acheuleans conducted their initial food preparation at some distance from their now-uncovered occupation spots, the consumption areas (= circular concentrations) enclosed.

This latter presumption makes more sense now, perhaps because of the opportunity that meanwhile Hay's volume (1976) has been published. All those floors that contain elongated concentrations are of pre-disconformity sites. These are from periods marked by an expanded lake, damper, closer, and denser vegetation, and occupation confined strictly to the lake margin facies. This was also the mesoenvironment where the most obtainable animals occurred. It seems that all dietary activities had to be, and in fact were, performed at near proximity to each other (killing, butchering, and eating); this may be the reason for the occurrence of elongated along-side circular concentrations. In the elongated ones a great proportion of the tools are large, heavy-duty, and usually, though not always, made of lava.

All other floors with circular concentrations only (= consumption areas) belong in post-disconformity times. The landscape then was more open, more grassy, the lake much smaller and at a greater distance to the west. The climate was drier, and both savanna animals and hominids were more dispersed along stream banks and on small pond margins. Hunting required more movement and wider territories. Food could no longer be brought through the 'door' adjoining the 'kitchen' (elongated concentration) and the 'dining room' (circular concentration). The animal thus had to be killed away from the home base, its edible and useful parts 'schlepped' in. The kill site was left unprotected and unwatched, with only a few remaining signs of the activity (several large, heavy-duty tools, a few other artifacts, and some heavy stones strewn around): scavengers probably quickly obliterated whatever was neglected by the hunters.

143

Since final consumption is presumed to have been mostly accomplished with the aid of small tools and debitage (Ohel 1977), and since small, light-duty tools, light-duty utilized flakes, and debitage are made almost exclusively of q/zite (Leakey 1971), sour interpretation suggests the representation of grouping III (post-disconformity) shown on Figure 12. Even if these people did use more large, heavy-duty lava tools than discovered (Figure 12:A), our chances of allocating them in some kind of reliable context seem to be extremely slim. In sum, what is now seen as a cognizant preference of selection might be nothing more than an 'archaeological accident' (and see also Schiffer 1976). This, too, remains a hypothesis to be tested.

SUMMARY AND CONCLUSION

The heuristic flow model (Figure 4) can now be tentatively filled in (Figure 13) serving as both summary and conclusion. While this article attempts only to outline several approaches that may or may not be elaborated in the future, it supports my belief that spatial analysis bears promise as an analytical tool for archaeological phenomena not only from later periods such as the Roman, the Neolithic, or the Mesolithic, but from earlier ones as well (cf. Foley 1977).

ACKNOWLEDGEMENTS

The first version was a course paper written for Brian Berry in 1974. I am grateful for his encouragement. Karl Butzer and Leslie Freeman read early drafts. Special thanks are due to Beth Dillingham for urging me to rewrite as well as for editing the manuscript. I alone, however, am responsible for all shortcomings. The final version was written while enjoying the hospitality of the University of Cincinnati, Department of Anthropology, as a Taft Post-doctoral Fellow during 1977-78.

REFERENCES

Berry, Brian J.L. 1967. *Geography of Market Centers and Retail Distribution.* Englewood Cliffs: Prentice-Hall.

Berry, Brian J.L. 1972. Deliberate change in spatial systems: goals, strategies, and their evolution. *SA Geogr. J.* 54:30-42.

Butzer, K.W. 1977. Environment, culture, and human evolution. *American Scientist* 65: 572-584.

Foley, R. 1977. Space energy: a method for analyzing habitat value and utilization in relation to archaeological sites. In: D.L.Clarke (ed.), *Spatial Archaeology* New York: Academic:163-187.

Freeman, L.G. 1975. By their works you shall know them: cultural developments in the Paleolithic. In: K.Eibel-Eibesfeld (ed.), *Hominisation und Verhalten,* Stuttgart: Fischer:234-261.

Gould, P. & R.White 1974. *Mental Maps.* Harmondsworth: Penguin Books.

Haggett, P. 1966. *Locational Analysis in Human Geography.* New York: St. Martin's.

Hay, R.L. 1976. *Geology of the Olduvai Gorge.* Univ. of California Press, Berkeley.

Hodder, I. 1977. Some new directions in spatial analysis of archaeological data at the regional scale (macro). In: D.L.Clarke (ed.), *Spatial Archaeology,* Acad. Press, London:223-351.

Hodder, I. & C.Orton 1976. *Spatial Analysis in Archaeology.* Cambr. Univ. Press, Cambridge.

Jochim, M.A. 1976. *Hunter-Gatherer Subsistence and Settlement.* Acad. Press, New York.

Larson, P. 1975. Trend analysis in archaeology: a preliminary study of intrasite patterning. *Norwegian Archaeological Review* 8:75-80.

Leakey, M.D. 1971. *Olduvai Gorge, Vol. 3. Excavations in Beds I and II, 1960-1963.* Cambr. Univ. Press, Cambridge.

Leakey, M.D. 1975. Cultural patterns in the Olduvai sequence. In: K.W.Butzer & G.Ll. Isaac (eds.), *After the Australopithecines,* Mouton, The Hague:477-493.

Leakey, M.D. 1976. The early stone industries of Olduvai Gorge. In: D.J.Clark & G.Ll. Isaac (eds.), *Les Plus Anciennes Industries en Afrique.* Prétirage, IXe Congrès UISPP, Nice:24-41.

Lynch, K. 1960. *The Image of the City.* MIT Press, Cambridge, Mass.

Lynch, K. *What Time is This Place?* MIT Press, Cambridge, Mass.

Ohel, M.Y. 1977. Patterned concentrations on living floors at Olduvai, Beds I and II: experimental study. *Journal of Field Archaeology* 4:423-433.

Plog, S. 1976. Measurement of prehistoric interaction between communities. In: K.V. Flannery (ed.), *The Early Mesoamerican Village,* Academic Press, New York:255-272.

Reidhead, V.A. 1979. Linear programming models in archaeology. *Ann. Rev. Anthrop.* 8:543-578.

Schiffer, M.B. 1976. *Behavioral Archeology.* Academic Press, New York.

Speth, J.D. & D.D.Davis 1976. Seasonal variability in early hominid predation. *Science* 192:441-445.

Stiles, D. 1979. Early Acheulian and Developed Oldowan. *Current Anthrop.* 20:126-129.

Stiles, D.N., R.L.Hay & J.R.O'Neil 1974. The MNK chert factory site, Olduvai Gorge. *World Archaeology* 5:285-308.

Thomas, D.H. 1976. *Figuring Anthropology.* Holt, New York.

Wilson, A.G. 1971. *Entropy in Urban and Regional Modelling.* Pion, London.

LATERAL FACIES ANALYSIS OF THE KOOBI FORA FORMATION, NORTHERN KENYA: SUMMARY OF RESEARCH OBJECTIVES AND RESULTS

ANNA K. BEHRENSMEYER
Yale University, New Haven, Conn., USA

LÉO F. LAPORTE
University of California, Santa Cruz, USA

This project was designed to build upon the previous ten years of research in the geology, paleontology and paleoanthropology of the Koobi Fora Formation, well-known for its record of Plio-Pleistocene hominids and archeology. The overall goal of the work was to integrate the many different kinds of evidence available from the rock and fossil record and derive a paleogeographic and paleoecologic reconstruction for a single, time-synchronous interval within the formation. The interval chosen, referred to as 'KOLM', comprises 12-17 meters of tuffaceous clastic sediment and is widely exposed over the East Lake Turkana area. It includes the local reference units known as the Koobi Fora, Okote and Lower/Middle Tuffs, which are believed to represent approximately the same time period at 1.5 ± 0.1 my (Findlater 1978). The KOLM includes fluvial, deltaic and shallow lacustrine sedimentary environments, abundant invertebrate and vertebrate fossils, and archeological sites of the Karari Industry (Harris 1978).

The results of this project are appearing in a series of publicatons, some of which are listed in the end of this report. Our purpose in this brief summary is simply to outline the objectives, results and personnel of the study and provide references for those interested in further information.

Field work was carried out during 1978-79, with follow-up studies continuing into 1981. Specific objectives included:

1) Characterization of the vertical and lateral patterns in lithology and sedimentary structures.

2) Documentation of all organic remains and traces in relation to the lithofacies.

3) Detailed examination of the taphonomy of vertebrate fossils, especially mammal and fish remains.

4) Paleogeographic reconstruction of the East Lake Turkana area for the time 1.5 ± 0.1 my BP, including characterization of the lake-land system and the paleoecology of vertebrate and invertebrate groups.

5) Concordant studies of modern environments and faunas, as needed to help in understanding of the Pleistocene record.

The research has been carried out by the authors, in collaboration with Mr Andrew Cohen (Department of Geology, University of California, Davis, Calif. 95616), Ms Hilde Schwartz (Earth Sciences Board, University of California, Santa Cruz, Calif. 95064) and Dr Jeffery Mount (Department of Geology, University of California, Davis, Calif. 95616), and with the assistance of Mr Mutete Nume and other technicians of the National Museums of Kenya, Nairobi. Much of the work is still in progress. The following list summarizes the nature of our data base and the person or persons primarily responsible for each part of it:

Sedimentology	– stratigraphic sections, lithologic samples for three major outcrop areas (Laporte, Mount)
	– microstratigraphy of excavations and geological step trenches (Behrensmeyer, Schwartz)
	– lateral marker unit mapping (Behrensmeyer, Cohen)
Paleontology	– bivalves, gastropods, sponge spicules, diatoms, grass phytoliths (Laporte)
	– ostracodes (Cohen; see this volume)
	– root casts (Cohen)
	– vertebrate tracks and trails (Laporte, Behrensmeyer)
	– fossil fish, taxonomy, taphonomy and paleoecology (Schwartz)
	– fossil mammals, taphonomy and paleoecology (Behrensmeyer)
	– reptiles and birds (Schwartz and Behrensmeyer)
Modern analogues	– controls on benthic faunas of modern East African lakes (Cohen)
	– bioturbation on recent land surfaces by large vertebrates (Laporte)
	– bone destruction and burial in modern ecosystems (Behrensmeyer)

From laboratory analysis of sediment samples, textural classifications of individual bedding units for each of the major study areas of the KOLM are now available. Careful processing of some of these samples led to the discovery of the preserved silicious microfossils, which may hold important paleoenvironmental information. Ostracodes are the most abundant and best preserved of the invertebrate organisms, and are making fundamental contributions to the reconstruction of paleo-Lake Turkana. Analysis of paleosols and soil carbonate horizons is being undertaken by Mount. Study of the fish paleontology by Schwartz holds great promise for contributing a previously neglected component of Pleistocene history. Detailed investigation of in situ as well as surface bone assemblages by Behrensmeyer will contribute interesting comparative data with respect to the question of archeological site formation. An unexpected new source of paleoecological information turned up during geologic trenching in the form of numerous bedding planes with preserved tracks of large vertebrates. Among these was the trackway of an early hominid, the second to be uncovered in East Africa (Behrensmeyer & Laporte, in press).

Although we have only begun integrating the sedimentological and paleontological information, a few preliminary statements can be made concerning the paleogeography and paleoecology of the KOLM interval. The two deltaic systems originally recognised by Findlater (1978) differ in terms of the preserved sub-environments and in some aspects of the faunas, although an overall similarity in lithology and mammal taxa is also apparent. Local ecological factors appear to be important in causing these differences in our time-synchronous unit. Mammal faunas vary most strongly across local facies boundaries (e.g. between contemporaneous channel and floodplain deposits). Another result from our study is that bone patches within a low density 'background' seem to be a typical pattern for fossil occurrences in both land surface and channel contexts. The bone patches may occur due to a number of taphonomic processes, only one of which is hominid activity. In stratigraphic sequences, bone horizons occur most frequently at points of marked change in sedimentation, especially at contacts between paleo-land surfaces and transgressive lacustrine deposits.

We have greatly appreciated the support and co-operation of the Government of Kenya and the National Museums of Kenya in our research on the Koobi Fora Formation. Mr Richard Leakey and Dr Glynn Ll. Isaac, acting as directors of the Koobi Fora Research Project and as research colleagues, have provided much helpful advice and assistance. We thank Dr Meave Leakey and Mr Mahmood Raza for their help in field work and contributions to discussion. The work has been supported by the National Science Foundation, Grant #EAR77-23149.

REFERENCES

Behrensmeyer, A.K. & Léo F.Laporte in press. Footprints of a Pleistocene hominid in northern Kenya. *Nature.*
Cohen, A. this volume. Paleolimnological research at Lake Turkana, Kenya.
Findlater, I.C. 1978. Stratigraphy. In: M.G.Leakey & R.E.Leakey (eds.), *The Koobi Fora Research Project, Volume 1: The fossil hominids and an introduction to their context.* Oxford, Clarendon Press: 14-31.
Harris, J.W.K. 1978. *The Karari Industry, its place in East African prehistory.* PhD dissert., Dept. of Anthropology, Univ. California, Berkeley.

Publications and manuscripts not referenced in the text

Published
Cohen, A.S. 1979a. Environmental significance of fossil root casts from Koobi Fora Fm. (Pliocene-Pleistocene), East Turkana, Kenya (Abstract). *Amer. Assoc. Petroleum Geologists Bull.* 63: 434pp.
Cohen, A.S. 1979b. Evolution of Late Tertiary-Modern benthic communities in Lake Turkana Basin, northern Kenya (Abstract). *Geol. Soc. America Abstract Volume, Cordilleran Section:* 72-73.
Cohen, A.S., K.E.Higgins & N.G.Dickinson 1979. The paleolimnologic history of Lake

Turkana (Abstract). *Soc. Inter. Limnol. Workshop on African Lakes.* UN Environmental Progr., Nairobi.

Cohen, A.S. 1980. Benthic faunal diversity regulation in Lake Turkana, Kenya (Abstract). *Geol. Soc. Amer. Ann. Mtgs.,* Atlanta, GA.

Laporte, L.F. & A.K.Behrensmeyer 1979. Tracks and substrate reworking by terrestrial vertebrates in Quaternary sediments in Kenya (Abstract). *Amer. Assoc. Petroleum Geologists Bull.* 63: 485-486.

In press

Laporte, L.F. & A.K.Behrensmeyer 1980. Tracks and substrate reworking by terrestrial vertebrates in Quaternary sediments in Kenya. *Jour. Sed. Petrology:* 50(6) (December).

In preparation

Cohen, A.S. Lacustrine paleochemical interpretations based on East and South African ostracodes. For submission to *Limnology and Oceanography.*

NAMIB IV AND THE ACHEULEAN TECHNO-COMPLEX IN THE CENTRAL NAMIB DESERT (SOUTH WEST AFRICA)

MYRA SHACKLEY

Department of Archaeology, University of Leicester, UK

SUMMARY

The author is at present engaged in a systematic survey of the archaeology of the central Namib desert, and in 1978 a preliminary reconnaissance was made at the Acheulean site of Namib IV, located in an interdune flat in the linear sand dunes (Figure 1). A further visit was made to the site in 1980 which produced more details about the artefacts and further faunal remains, including the mid-Pleistocene *Elephas recki*. During the 1980 season, a further four sites were located (Xmaspan, Tsondab route, Narabeb West and Zebravlei) and a new assemblage studied from Narabeb where Acheulean material had already been reported by Seely & Sandelowsky (1974). Since less than a month has passed since the end of this 1980 field season, it is not yet possible to describe the characteristics of these Acheulean technocomplex sites in detail, but they will provide the beginnings of a framework for an understanding of the Pleistocene occupation of the Namib, which at present is seen to begin with the kill and butchery site of Namib IV and its associated fauna (Shackley 1980).

1. INTRODUCTION

The surface assemblage of artefacts and faunal remains called Namib IV is located in an interdune flat where different generations of fossil calcretes have been exposed by sand shift. The material has recently weathered out of a dark red calcrete, traces of which may be seen on the ventral surfaces of much of the material. It seems likely that the site is the remains of a kill and butchery area near the margin of a former waterhole which, after it had dried, was buried by sand shift. Post-Pleistocene dune movements then exposed the calcrete and its included material initiating a new weathering cycle, resulting in the present surface distribution of material. Such a sequence of events, with local variations, is responsible for the present configuration of most of

151

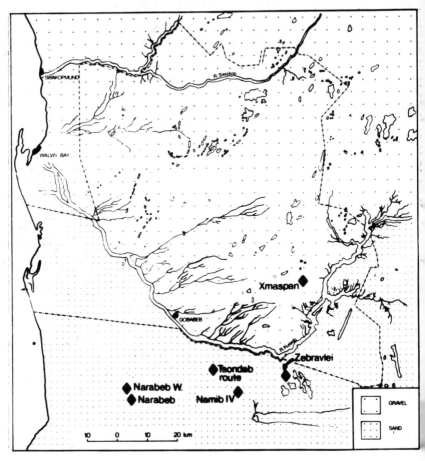

Figure 1. Surface sites provisionally assessed as belonging to the Acheulean techno-complex in the central Namib Desert, south west Africa.

the archaeological sites of the Namib, and new material is constantly being revealed by sand movements.

In 1978 a grid was set out across the interdune flat and a statistical and typological analysis of the contents of one randomly-selected grid square (2,500 m²) was undertaken. The 394 implements studied were supplemented by a further assemblage of 107 artefacts considered in the 1980 field season, which also produced extensive new faunal remains which are at present awaiting analysis.

152

2. THE ARTEFACTS

The 1978 and 1980 artefact assemblages total 501 pieces, 57 % of which could be classified as 'heavy duty tools' (choppers, cleavers and handaxes). The 43 % 'light duty tools' included points, scrapers, flakes and blades but no evidence of the prepared-core technique which has been found at other Acheulean sites. The raw materials were very varied and included quartzites (yellow, orange, pink and grey) with lesser quantities of dolomite, dolomitic marble and diamictite; all probably obtained as water rolled cobbles from the bed of the ephemeral river Kuiseb 8 km to the north. There was a definite relationship between raw material and artefact type. The large cleavers, for example, were almost exclusively made from quartzites and dolomite was used for hammerstones. The smaller flake tools were occasionally made from quartz (although this raw material was not so common here as at the other sites), obtained locally as vein quartz from outcrops of mica schist.

Various technological features of the Namib IV assemblage make it particularly distinctive. Most of the cleavers and a few of the handaxes were made by splitting rounded quartzite cobbles of suitable size and trimming the resulting large thick flakes to the desired shape. In the case of flake cleavers this trimming was often minimal (Figure 2) so that the ventral surface of the tool had but a few small trimming flakes removed from it and the dorsal surface consisted entirely of cortex. This approach, involving the use of as few blows as was consistent with the production of a serviceable edge and the required shape is not only related to economy of effort but to the extreme difficulty of working quartzite, a fact not sufficiently appreciated by many archaeologists. It requires a great deal of force to split a quartzite pebble and this is easiest when the raw material has been heated by the sun. Full advantage must also be taken of any lines of weakness in the pebble but a 'block on block' technique is the only way of initially splitting it, although a hard hammer may be used for shaping. Retouching quartzite is not the easy matter that it is with softer rocks such as flint, a point which probably accounts for the relatively low frequencies of retouched tools in Namibian sites and the small number of flake scars on bifaces. At Namib IV more than ten flake scars per side is unusual and over 50 % of the bifacial tools have less than seven. The method of production outlined above also accounts for the high frequency of handaxes and cleavers made on flakes (84 % of cleavers), often side struck. These characteristic Namib IV divergent cleavers (Figure 2(1)) have only been found at one other Acheulean site in the area (Tsondab route) where the same manufacturing method has been used.

Another feature of interest is the occurrence of 'chisel-ended' handaxes (called convergent cleavers by Kleindienst 1962) although the bifacial component of the industry also includes more 'classic' pointed ovate handaxes (Figure 2(4)) with a notable tendency towards plano-convexity (again a function of the method of manufacture). Cleavers were nearly three times as common as other bifacial tools (37 % cleavers, 12 % handaxes) and cores

153

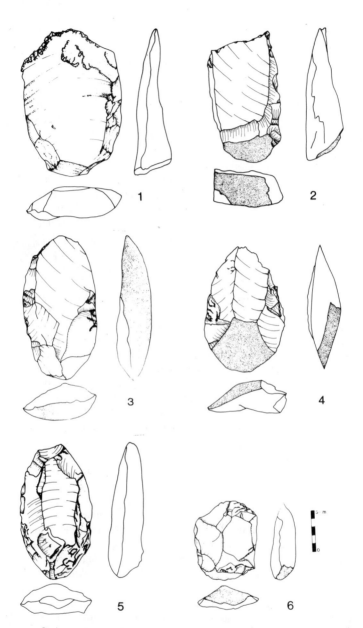

Figure 2. Implements from Namib IV. 1-2 – Divergent quartzite flake cleavers; 3 – Flake cleaver made from a split ovate quartzite pebble; 4 – Pointed handaxe with cortex butt made from a split ovate quartzite pebble; 5 – Irregular ovate handaxe made from poor-quality quartzite core; 6 – Small rectangular quartzite handaxe made on a thick flake.

formed only 2 % of the whole assemblage. This is also a function of the type of raw material being used but the lack of small flake cores is perhaps surprising in view of the comparatively high number of scrapers (7 %) and points (23 %). At Namib IV, as at other sites in the area, the continuum which exists between raw material, hammerstone and finished product is very striking and seems likely to require a revision of technological nomenclature. The transportation of a lump of rock to the site qualifies it as a manuport and as raw material for tool production. If one cobble is struck against another until a flake is removed one then has an arbitrary division into hammerstone and core, and as more flakes are removed an implement which may be described as either a core or a chopper is often produced; it certainly functioned as both. By far the majority of cores from all Namibian sites show signs of having been used as choppers so what is the distinction? All the artefacts from Namib IV whether bifacial, core, chopper or flake also show signs of heavy use-wear which is another characteristic of the Namibian Acheulean industries.

Isaac (1969) suggested that the manufacture of bifacial tools from large flakes may be crucial in differentiating Acheulean industries from Oldowan. The Namib Acheulean industries do not, however, all contain bifacies made on flakes and at the other sites (Figure 1) this technique may occur on less than 10 % of the handaxes. In the authors' opinion the manufacture of bifacial tools on flakes is largely governed by the nature of the raw material and is more likely to occur where, as at Namib IV, the artefacts are made from pebbles. Some techniques do not occur at Namib IV but are found in other sites in the area, for example at Narabeb West where 50 % of the handaxes are ovates with a pronounced S-twist, rare in the African Acheulean technocomplex. At Narabeb, Narabeb West and Zebravlei the industries include a Levallois component together with prepared single-platform cores, and in some cases points represent over 50 % of the 'light duty tools', approaching the figures for middle stone age industries.

The closest typological parallels for Namib IV are undoubtedly to be found in East Africa, at the sites of Olorgesailie (Isaac 1977), Kilombe (Gowlett 1978) and PDK and HEB 1 and 3 at Olduvai Bed IV (Leakey 1975). It could be argued (Shackley 1980) that the thick, crude Namib IV bifacies with their low number of flake scars are even earlier than the East African sites, but in order to make such a comparison detailed study of the comparative technologies involved must be undertaken. There is, however, no doubt that the Namib IV industry must be placed firmly within the Acheulean technocomplex. Namib IV contrasts with the East African sites in its dominance of cleavers over handaxes, but if one is manufacturing a bifacial tool from a spherical pebble the cleaver is the simpler form. The absence of picks and the presence of sub-rectangular cores, spheroids and large numbers of flake tools is similar to the East African material. It is interesting to note that cleavers do not dominate the assemblage at any of the other Namibian Acheulean sites, being present only at Tsondab route. Certain features of the other assemblages (for example, the use of the prepared core technique) suggest that they may be later than Namib IV.

155

Preliminary readings from Kilombe (Dagley et al. 1978) suggest that the site antedates the Brunhes-Matuyama transition and is therefore >700 000 years old. Isaac (1977) is of the opinion that Olorgesailie may date to >400 yr BP, and it is suggested on typological grounds that Namib IV must be at least of similar antiquity.

3. THE FAUNA

The faunal remains were mineralised, fragmented and found in the same area as the implements. The 1978 faunal assemblage was identified by Professor R.G.Klein but the 1980 finds have not yet been analysed. The fauna included the remains of *Elephas recki* together with one indeterminate alcelaphine antelope and one other medium-sized antelope. The *Elephas* remains fortunately included tooth fragments whose characteristics (thin enamel, hypsodonty and tight enamel folding) suggested that a late stage of *E.recki* was represented, probably comparable with Maglio's Stage 4 (Maglio 1973) as represented at Olduvai Bed IV. This confirms the mid-Pleistocene date for Namib IV and is especially interesting in view of similar finds from the Southern Kalahari site of Kathu Pan where *E.recki* remains have also been found in association with an Acheulean industry (Klein, personal communication).

4. THE SITE AND ITS IMPLICATIONS

It seems likely that the calcrete deposits are connected with the former presence of surface water, probably a small ephemeral lake. It was previously suggested (Shackley 1980) that the presence of elephant indicated that in mid-Pleistocene times the Namib was receiving at least sufficient precipitation to support savannah grassland but it is equally possible, though in the writer's view less likely, that the elephants were only migrating through the area. The repeated association of Acheulean artefacts and interdune calcretes in the Namib undoubtedly represents a hunting pattern of kill and butchery sites near waterholes situated not too far from suitable sources of raw material for toolmaking. Namib IV was probably both a kill, butchery and workshop site although it seems unlikely that it was ever occupied for any substantial length of time. The presence of game at such waterholes could be predicted and the chances of a kill were sufficiently high to justify the transport of heavy raw materials from a distance away. The paucity of bone refuse from Namib IV makes it impossible to come to any conclusion about the number and distribution of the animals which were killed; if indeed they were killed by man and not merely scavenged from the kills of the large carnivores who would have represented intense competition and probably meant that the occupants of the sites were unlikely to remain in the vicinity of their kills for very long (Klein 1978). The availability of water and food even on a seasonal basis, would no doubt mean that Namib IV was a location which was visited on more than one occasion. This is, indeed, the general pattern of occupation of all the Namib sites, which seem to have been utilised for their particular resources by different groups of people visiting the area over a period of many years, but not making permanent camps there. Such an occupation pattern makes the lot of the typologist an especially hard one since the visits may be spaced in time yet by the archaeological 'telescoping' effect appear contemporary. It would be possible to envisage a utilisation pattern which involved several groups of people in the central Namib at the same time but with different tool-making traditions, as well as groups of people who came at different times with the same artefact types. Associated faunas are clearly going to be of the greatest importance here and much more detailed ground survey and artefact analysis is needed before even a preliminary model can be obtained. One thing is clear, and that is that the possibility of finding assemblages with virtually the same artefact component types and percentages is remote. Many of the early hypotheses have to be revised; for example the theory that Acheulean material is never found north of the Kuiseb (Seely & Sandelowsky 1974) which has been confounded by the discovery of the Xmaspan site. The six sites shown on Figure 1 are at present being attributed to the Acheulean technocomplex on typological grounds, but an examination of Namib IV has shown that the nature of the raw material and methods of manufacture may to a certain extent control implement typology. It seems likely that not only will the structure of the Acheulean assemblages of the Namib turn out to be incredibly varied but that the conventional division between the Acheulean and later technocomplexes may become meaningless here due to the nature of the use pattern of the area.

In conclusion, it may be said that Namib IV has provided a first peg in the framework for the Pleistocene occupation of the Namib by its association of artefacts and fauna with clear parallels from other African sites. It is to be hoped that future work will enable us to establish a chronology for the different tool-making traditions of the central Namib, the detailed analysis of the five new 1980 assemblages shown in Figure 1 being a good starting point.

ACKNOWLEDGEMENTS

The 1978 fieldwork was financed by grants from the Swan Fund (University of Oxford) and a Suzette Taylor Travelling Scholarship (Lady Margaret Hall, Oxford). In 1980 financial support was obtained from the British Academy and the Boise Fund (University of Oxford). I am most grateful to all the following individuals and institutions for their assistance and permissions: Richard Klein, Ann Woods, Mary Seely and the staff of the Desert Research Unit (Gobabeb) and the Division of Nature Conservation and Tourism (South West Africa).

REFERENCES

Dagley, P., A.E.Mussett & H.C.Palmer 1978. Preliminary observations on the palaeomagnetic stratigraphy of the area west of Lake Baringo, Kenya. In: W.W.Bishop (ed.), *Geological Background to Fossil Man.* Scottish Academic Press, Edinburgh:225-237.

Gowlett, J.A. 1978. Kilombe – an Acheulian site complex in Kenya. In: W.W.Bishop (ed.), *Geological Background to Fossil Man.* Scottish Academic Press, Edinburgh:337-361.

Isaac, G.Ll. 1969. Studies of early cultures in east Africa. *World Archaeology* 1:1-28.

Isaac, G.Ll. 1977. *Olorgesailie,* University of Chicago Press, Chicago & London, 272pp.

Klein, R.J. 1978. The fauna and Overall Interpretation of the 'Cutting 10' Acheulean site at Elandsfontein (Hopefield), Southwestern Cape Province, South Africa. *Quat. Res.* 10:69-83.

Kleindienst, M.R. 1962. Components of the East African Acheulean assemblage: analytical approach. In: *Actes IV Congrès Panafr. Préhist. et du Quat. Léopoldville, 1959:*81-111.

Leakey, M.D. 1975. Cultural Patterns in the Olduvai Sequence. In: K.W.Butzer & G.Ll. Isaac (eds.), *After the Australopithecines: Stratigraphy, Ecology and Cultural Change in the Middle Pleistocene.* Mouton, The Hague:477-494.

Maglio, V.J. 1973. Origin and Evolution of the Elephantidae. *Trans. Amer. Philos. Soc.* 63:1-149.

Seely, M.K. & B.H.Sandelowsky 1974. Dating the Regression of a River's End Point. *S. Afric. Archaeol. Bull.* Goodwin Series 2:61-64.

Shackley, M.L. 1980. An Acheulean industry with *Elephas recki* fauna from Namib IV, South West Africa (Namibia). *Nature* 284. (5754):340-1.

158

SHORT COMMUNICATIONS

IONIUM DATING OF PEAT: A NOTE

J. C. VOGEL

Natural Isotopes Division, NPRL, CSIR, Pretoria, South Africa

Ionium dating or uranium series disequilibrium dating has been applied successfully to calcareous lake deposits (Kaufman & Broecker 1965, Kaufman
1971) and speleothems (Thompson *et al.* 1975, Ku 1976). The method is
based on the observation that thorium and thus also ionium (^{230}Th) is rapidly
removed from solution by adsorption to solid surfaces and that natural waters
therefore contain dissolved uranium, but virtually none of the decay product,
^{230}Th. When dissolved uranium is co-precipitated with calcium in fresh water
limestone deposits, the concentration of the ionium daughter starts to
increase with a doubling time of 75 200 years until radioactive equilibrium is
again established.

It has been found recently that, under certain circumstances, the dating
method can also be applied to peat deposits (Vogel & Kronfeld 1980). Organic matter, and specifically peat, is well-known to strongly adsorb uranium
from solution; 'enrichment factors' (absorption ratios) of 10 000 have, for
instance, been measured on peat (Szalay 1954). In consequence of this the
surface layers in a peat bog will effectively remove uranium from aqueous
solution. When this surface layer becomes buried under new accumulations
of peat, it will be shielded from further contact with uranium. Provided, therefore, the absorption capacity of the surface layer is high enough, uranium will
not penetrate the deposit to any marked degree and the underlying layers will
constitute a closed system. The gradual accumulation of the daughter isotope,
ionium, by radioactive decay of the uranium will then provide a measure for
the time that has elapsed since the original formation.

If the peat is not purely organic and contains significant amounts of inorganic detrital material, this 'ash' will normally have a certain amount of thorium adsorbed to it which will constitute a source of error since this intro-
duces an essentially unknown quantity of initial ^{230}Th. If, however, the
uranium content of the peat is sufficiently high, the uncertainty in the amount
of initial ^{230}Th will constitute only a minor increase in the error attaching to
the calculated age.

The evidence gained thus far suggests that low moor peats can be dated
successfully by the method. On the other hand, a raised bog in the Nether-

Table 1. Ionium dates obtained on peat samples

Sample	U conc. (ppm)	C[14] age* (years BP)	Th[230] age (years BP)
U-190 Tenagi Philippon 1, N.Greece	21.5	8 040 ± 50	8 380 ± 850
U-178 Peelo, Netherlands	13.9	34 620 ± 450	36 200 ± 4 300
U-181 Zell am Inn, Bavaria	10.5	>66 640	76 100 ± 4 100
U-189 Grossweil 2, Bavaria	6.7	early Würm	85 900 ± 6 900
U-18 Grossweil 1 wood	0.16	late Riss-Würm	93 000 ± 9 000
U-194 Grossweil 1	14.3	Interglacial	116 000 ± 7 800

* C.[14] ages calculated with a half-life of 5 730 years. Dates and details are listed in Radiocarbon 9 (1967) p.63-106 and 14 (1972) p.6-110.

lands which presumably only came into contact with rain water, does not contain sufficient uranium for the calculation of ages. Some gyttja samples were also not usable for the same reason. Examples of results obtained on peat samples from Europe, some of which have also been dated by radiocarbon, is given in the table. The correspondence between the two dating methods is sufficiently close to warrant further investigation of the method, especially since ionium dating can provide absolute ages for palynologically investigated peat layers back to 300 000 years BP.

REFERENCES

Kaufman, A. & W.Broecker 1965. Comparison of Th[230] and C[14] ages for carbonate materials from Lakes Lahontan and Bonneville. *J. Geophys. Res.* 70: 4039-4054.
Kaufman, A. 1971. U-series dating of Dead Sea basin carbonates. *Geochim. et Cosmochim. Acta* 53: 1269-1281.
Thompson, P., D.C.Ford & H.P.Schwarcz 1975. [234]U/[238]U ratios in limestone cave seepage waters and speleothems from West Virginia. *Geochim. et Cosmochim. Acta* 39: 661.
Ku, T.L. 1976. The uranium series methods of age determination. *Ann. Rev. Earth Plan. Sciences* 4: 347-379.
Szalay, A. 1958. The significance of humus in the geochemical enrichment of uranium. In: *Peaceful uses of atomic energy.* 2nd UN Intern. Conf., Proc. Geneva, vol.2: 182-186.
Vogel, J.C. & J.Kronfeld 1980. A new method for dating peat. *S.Afr. J. Sci.,* Dec. 1980.

DROUGHT PREDICTION IN THE SAHEL?

In a paper to be published in ASEQUA[1] 7 Bulletin in early 1981 by Hugues Faure, from CNRS[2] and University of Marseille-Luminy (France), and Jean-Yves Gac, a geologist at ORSTOM[3] in Dakar (Sénégal), it is shown that the Sénégal river discharge, measured since the beginning of this century shows a cyclic pattern.

The drastic sahelian drought has a 31 year return cycle (around 1913, 1941, 1975). But the anomalous years of river discharge (25% above or under the mean) show a 10.4 year cycle almost identical to the sunspot cycle during the same period.

An extrapolation of the curves shows that the present drought should end around 1985. But these better conditions could probably cease at about 2005 with the incoming of a new drought. It is possible to prepare a fight against this future drought by using the period of propitious conditions.

Studies are going on at CNRS and ORSTOM in connection with Universities of the Sahelian countries and various organisms to extend research on cycles of different nature. Various time scales are considered during Quaternary and recent periods. The project is named CYCLARID.

H.Faure

1. ASEQUA – Association Scientifique pour l'Etude du Quaternaire (Université de Dakar, Sénégal).
2. CNRS – Centre National de la Recherche Scientifique.
3. ORSTOM – Office de la Recherche Scientifique et Technique Outre-Mer.

PRÉDICTION D'UN FUTUR CYCLE DE
SÉCHERESSE AU SAHEL?

Dans un article qui doit paraître dans le prochain numéro du bulletin de l'ASEQUA[1] rédigé par Hugues Faure, du CNRS, Faculté des Sciences de Marseille-Luminy (France) et Jean-Yves Gac, de l'Office de la Recherche Scientifique et Technique Outre-Mer de Dakar (Sénégal) il est montré que le débit annuel du fleuve Sénégal mesuré depuis le début de ce siècle présente des variations cycliques.

Les grandes sécheresses du Sahel bien marquées dans les courbes de débit du fleuve reviennent environ tous les 31 ans (autour de 1913, 1941, 1975). Mais les années anormales pour lesquelles le débit est supérieur ou inférieur de 25 % par rapport à la moyenne apparaissent tous les 10,4 ans presqu'en coïncidence avec les années de minimum d'activité solaire. L'extrapolation des courbes permet de prédire que la sécheresse extrême du Sahel se terminera vers 1985 par un retour à des conditions meilleures qui cesseront vers l'an 2005 pour faire place à un nouveau cycle aride. On conçoit qu'il faut préparer la lutte dès maintenant pour avoir des chances de succès lors du futur cycle aride.

Les études se poursuivent au CNRS (Laboratoire de Géologie du Quaternaire) et à l'ORSTOM en liaison avec les Universités des pays du Sahel (IFAN) et plusieurs organismes afin d'étendre l'étude des cycles dans des domaines variés. Les échelles de temps considérées sont essentiellement celles du Quaternaire et du Récent. Le projet s'intitule CYCLARID.

H.Faure

---

1. ASEQUA – Association Scientifique pour l'Etude du Quaternaire Africain. Département de Géologie, Université de Dakar, Dakar-Fann (Sénégal).

SYMPOSIUM ON METHODS OF RECONSTRUCTING PALEOCLIMATE

During the 4th Quaternary Conference, 18-20 May 1979, at York University, Toronto, speakers from across North America and Great Britain discussed methods of reconstructing paleoclimate. Particular attention focused on stratigraphic, pedologic, geomorphologic, isotopic, paleontologic, palynologic and archeologic evidence, as well as the degree of resolution possible in reconstructing the magnitude of climatic change. Some of the research topics discussed at the Symposium and in the proceedings volume are summarized as follows:

J.T.Andrews *et al.* (University of Colorado's Institute of Arctic and Alpine Research, Boulder) discussed the terrestrial and ocean paleoenvironment derived from morphostratigraphy, lithostratigraphy, faunal assemblages, and amino acid epimerization determined from shells in raised moraine sediments, which suggest warm inshore water conditions during the last glaciation.

Late Quaternary climatic changes in the Aleutian Islands, described by Robert F.Black (University of Connecticut), included details on ice-cap destruction and the advance and recession of alpine glaciers, changes in elevation of regional snowline, pollen stratigraphy, and tephra distribution.

Aleksis Dreimanis (University of Western Ontario) discussed climatic conditions deduced from the nature of glacial deposits. He discussed the paleoclimatic and paleoenvironmental implications of some genetic varieties of till (for example, meltout, sublimation and waterlain tills); some basal tills contain, as incorporated material, valuable records of those non-glacial intervals that preceded the deposition of the tills.

Alan V. and A.Morgan (University of Waterloo, Ontario) reviewed the history of paleoentomological research, and reported an abundance of organic detritus found in different stratigraphic horizons, including a rich assemblage of insects that are sensitive indicators of change in climate.

A paper, by D.Fisher & F.Koerner, was presented by S.Patterson (Polar Continental Shelf Project, Ottawa), which analyzed data retrieved from ice cores taken from the Devon Island ice cap. Power-spectral analysis is used to cluster oxygen-isotope ratios to provide a detailed paleotemperature record since the end of the last glacial.

R.S.Harmon (Scottish Universities Research & Reactor Center, East Kilbride, Scotland) presented quantitative paleoclimatic information derived from stable-isotopic composition of speleothems in limestone caves. He stressed the use of stable- and chronologic-isotope data indicating that glacial climates occurred in North America from 275 000 to 230 000 years ago, from 190 000 to 135 000 years ago, and from 80 000 to 15 000 years ago.

William C.Mahaney (York University, Toronto) spoke about the use of buried and relict paleosols in the Rocky Mountains and East Africa as paleoclimatic indicators. Changes in climate induce new and sometimes different soil systems that survive burial and reveal paleoclimatic information; relict soils found at the surface often retain clay mineral species indicating a former wetter climate.

Richard G.Baker (University of Iowa) reported evidence for interglacial and interstadial conditions produced by pollen and plant macrofossil remains from three sections in Yellowstone Park older than about 70 000 years.

Abstracts and the field guide from the symposium can be obtained from the present author for $3.00.

William C.Mahaney
Dept Geography, Atkinson College
York University
4700 Keele Street, Downsview
Ontario M3J 2R7, Canada

The terrestrial ecology of Aldabra 1979, pp.263, by D.R.Stoddart & T.S.Westoll (eds.), The Royal Society, London.

The terrestrial ecology of Aldabra represent the collected papers of a Royal Society discussion meeting held on 16 and 17 March 1977 in London. (These papers were also published in the Philosophical Transactions of the Royal Society of London, Series B (No.1011) pp.1-263.

As a result of an announcement in 1966 that the British Government were considering the feasibility of constructing an air staging post on Aldabra, the Royal Society prepared a memorandum which stated that 'if development takes place on Aldabra the loss to science will be permanent'.

The remoteness of Aldabra makes it one of the least disturbed of the major coral atolls in the world, and as such the extensive endemic biota represent a unique site for scientific research. Since 1967 the Royal Society has administered a research station on Aldabra, and in 1980 this was handed over to the recently incorporated Seychelles Islands Foundation who are now attempting to raise funds from charitable donors to ensure the continuation of research and conservation.

This volume contains 23 papers which attempt to give a broad picture of terrestrial ecological research that has been carried out on Aldabra since 1967. Four papers deal with climate, soils and land forms, while a very interesting and important contribution is that of Taylor *et al.* who outline the changes that have occurred in the land faunas of Aldabra during the Pleistocene as a result of changing sea levels and total inundation. Four papers deal with various aspects of the Aldabran vegetation, and a further three discuss the insect, crustacean and other anthropod faunas of the terrestrial environment. Four papers deal with the land and avifauna including the rare endemic Aldabran brush warbler (*Nesillas aldabranus*).

The dominant feature of the Aldabran biota is the population of giant tortoises, which were common on the Mascarenes. Seychelles and other western Indian Ocean coral islands in the recent past. This population of up to 152 000 individuals has been the subject of extensive research for the last ten years. Five papers are included in this volume which describe the systematics of

Indian ocean tortoises, the history of their occurrence and extinction, their biomass and carrying capacity; reproduction and mortality. A further paper describes the nesting activity of the green turtle on Aldabra.

The appearance of this volume and the establishment of the Seychelles Islands Foundation leads us to hope that it will stimulate further international research effort on this unique atoll.

Malcolm Coe

REFERENCES

Bourn, D. & M.J.Coe 1979. Features of tortoise mortality and decomposition on Aldabra. *Phil. Trans. Roy. Soc. Lond.* B286: 189-193.

Coe,M. 1980. African mammals and savanna habitats. Int. Symp. *'Habitats and their influence on wildlife. Endangered Wildlife Trust,* Pretoria 3-4 July 1980: 83-109 (mimeographed).

Coe,M. 1980. African Wildlife Resources. In: M.E.Saute & B.A.Wilcox (eds.), *Conservation Biology.* Sinauer Ass., Massachusetts: 273-302.

Coe,M. 1980. The role of modern ecological studies in the reconstruction of palaeoenvironments in Sub-Saharan Africa. In: A.K.Behrensmeyer & A.P.Hill (eds.), *Fossils in the making.* Univ. Chicago Press: 55-71.

Coe,M.J., D.Bourn & I.R.Swingland 1979. The biomass, production and carrying capacity of giant tortoises on Aldabra. *Phil. Trans. Roy. Soc. Lond.* B: 163-176.

Swingland,I.R. 1977. Reproductive effect and life history strategy of the Aldabran giant tortoise. *Nature,* Lond. 269: 402-404.

Swingland,I.R. & M.J.Coe 1978. The natural regulation of giant tortoise populations on Aldabra Atoll: reproduction. *J. Zool. Lond.* 186: 285-309.

Swingland,I.R. & M.J.Coe 1979. The natural regulation of giant tortoise populations on Aldabra Atoll: recruitment. *Phil. Trans. Roy. Soc. Lond.* B286: 177-188.

The Sahara and the Nile. Quaternary environments and prehistoric occupation in northern Africa. 1980, 607pp., by M.A.J.Williams & H.Faure (eds.).

1 colour pl., 17 photos, Cloth. A.A.Balkema, Rotterdam. $49.50, £21.50 This impressive volume contains 22 contributions which give an up-to-date survey of former and ongoing studies, mainly on the late-Pleistocene and Holocene, of Northern Africa. The book is divided into three parts which are provided with concise introductions in French and in English. The articles written in these languages have abstracts in the alternate language. The content covers a very wide array of subjects, such as: geomorphology, historic and palaeo-climates, glacial and periglacial geology, lake levels, volcanism, tectonics, river alluvia, archaeology, etc.

The first 200 pages of the volume deal with the Sahara. A.T.Grove gives a brief introduction into the intricate Cenozoic history of the Sahara and the Nile. M.Mainguet, L.Canon and M.C.Chemin deal with the influence of the wind, the most active geomorphic agent in the Sahara, using satellite images.

Much attention is focussed on the climatic changes which occurred in the Sahara. As an introduction to this section M.R.Talbot describes the Climap results for the tropical eastern Atlantic Ocean and the atmospheric circulation during the last glacial maximum. B.Messerli, M.Winiger and P.Rognon give a survey of present and former glacial and periglacial altitudinal limits in the Atlas, the Saharan and East African mountains. It appears that the periglacial limits even in the Saharan uplands were surprisingly low during the last glacial maximum. A possible decrease in temperature in the Central Sahara of 10-14°C in winter sounds very substantial. It is further estimated that the lowering in temperature 18 000 yr BP in the East African mountains was 6-8°C, which concurs well with the fossil pollen data of 5,1-8,8°C. In another chapter, D. Livingstone, dealing with this drop in temperature, ignores this well-known pollen data. The Servants' in their description of the Quaternary history of the Chad basin imply drops in temperature during the time spans 26 000-20 000 BP and 12 000-7 500 BP and describe tropical diatom floras for the periods 20 000-18 000 BP and 7 000 BP to the present.

Much more information on changes in humidity than on temperature changes is available and is described by M.R.Talbot based on former dune and river activity, by the Servants on lake levels and by P.Rognon on the stratigraphy of the 'Ober-, Mittel- and Niederterrassen' in the Tibesti. L.Balout and C.Roubet base their climatic conclusions on archaeological evidence, while S.E.Nicholson gives an interesting review of historic data on climatic changes in the Sahara.

J.Maley deals with the late Cenozoic climatic changes in Africa and the origin of the Sahara. He explains the origin and age of the desert in relation to world wide decreases in temperature in Tertiary times and glaciations in both hemispheres.

The second part of the volume is devoted to the Nile. The intricate history of the Nile basin, which dates partly back to the Cretaceous and was strongly influenced by uplifts of the Ethiopian highlands and subsequent erosion and rifting, is described in a very clear way by the geologist couple F.M. and M.A.J.Williams. The strong influence of the old geological lineaments on the Nile drainage system and on the detailed course of the main rivers, has been studied in the field and with the aid of aerial and satellite pictures by D.A. Adamson and Mrs Williams.

K.W.Butzer gives a valuable summary of the alluvial history of the Nile Valley and the palaeoclimatological inferences which can be made from these processes. The study of the Quaternary history of the sediment loads of both the White and the Blue Nile by M.A.J.Williams and D.A.Adamson leads to an interesting explanation of the changes in water supply from the Ugandan headwaters. These findings are supported by D.A.Livingstone who reviews the climatic changes which might have occurred in the Nile headwaters, especially in Uganda. During the warming of the climate in Late-Glacial times the dry grasslands in East Africa were replaced by forest, lakes overflowed again and the Nile received much water from the Turkana-, Victoria- and Albert Lakes. In his overcritical evaluation of fossil pollen evidence and especially of pollen dispersal, he does not mention the new explanations this approach offers for his own results in the Ruwenzori complex.

The interesting studies by F.Gasse, P.Rognon and F.A.Street show that the Afar and Ethiopian lakes also expanded much in size during the warmer periods of the late Quaternary.

The geomorphic history of the isolated Jebel Marra volcano, a refugium of biogeographic importance, is described and reviewed by five geologists.

The third part of the book contains seven chapters on the prehistoric occupation of the Egyptian Sahara, the Nile Valley, the Sahel, the neolithic tradition and on domestication of cattle and plants. F.Wendorf and F.A.Hassan describe the Late Palaeolithic and Neolithic traditions of the Egyptian Sahara during the humid early Holocene and are of the opinion that the Neolithic people living round lakes in the present Western Desert were the first to cultivate plants in Africa 2 000 years after agriculture was initiated in the Near East. Hassan treating the same period for the Nile Valley in a separate chapter is of the opinion that the Neolithic people concentrated in the valley as the climate became drier and established farming communities about 1 000 years after the spreading of agriculture in the adjoining desert.

A.B.L.Stemler, discussing the origin of agriculture in Africa, mentions that archaeological evidence for cultivation does not go back further than 3 000 years when *Pennisetum* was grown in Dhar Tichitt in Mauritania. She suggests, however, that domesticated plants and animals might have been introduced into lower Egypt from the Near East during more humid early Middle Holocene times.

A description of the present life of nomads living in the West African Sahel by S.E.Smith shows how their movements depend entirely on the needs of their cattle. These nomads collect wild seeds of *Panicum* and *Eragrostis.*

A.B.Smith suggests that the domesticated cattle in the Sahara may have been of African origin. In the Sahel cattle breeding became very important at about 4 000-3 300 BP after the pastoralists had to move southward out of the Sahara because of the drought.

The last chapter of the book gives a masterly description of the prehistory of the Sahara and the Nile by J.Desmond Clark who emphasises that climatic pressure since 4 000 BP was an important factor for the introduction of agriculture.

The volume *The Sahara and the Nile* testifies to the great scientific activity, especially in the geological field, during the last decade. The book is edited with great care which is evident from the excellent summaries, the prefaces and the epilogue. Details on the authors and good indices make it a very useful book. The only criticism could be that some of the lettering on a number of maps and tables is hardly readable as authors very often produce large drawings without taking the necessary reduction in size into account.

The volume is highly recommended. No student of the geological and biological history of Africa can afford not to study the content of this book.

E.M.v.Z.B.

Florengeographische Untersuchungen im Raume der Sahara. Ein Beitrag zur pflanzengeographischen Differenzierung des nordafrikanischen Trockenraumes. 1978. Von Peter Frankenberg. Bonner Geographische Abhandlungen, Heft 58. Ferd. Dümmlers Verlag, Bonn.

Die vorliegende Arbeit führt zu einer floristischen Raumgliederung des Nordafrikanischen Trockenraumes der Sahara, wobei die Differenzierung von Holarktis und Paläotropis im Vordergrund steht.

Die Analyse beruht auf einer Auszählung von 4700 Pflanzenarten im Rahmen eines quadratischen Gitternetzes von 80 km Seitenlänge. Für 813 Gitterquadrate konnte nach Florenlisten und Literaturangaben eine ausreichende Information über die jeweils verkommenden Pflanzen gewonnen werden.

Die 4700 Pflanzenarten wurden nach ihrem Arealtypus 8 Florenelementen zugeordnet. Der relative Anteil der einzelnen Florenelemente pro Gitter an der jeweiligen Gesamtartenzahl ist in einer farbigen Karte der Arealtypenspektren wiedergegeben. Sie zeigt deutlich den kontinuierlichen Übergang von einer mediterranen Flora im Norden zu einer paläotropischen Flora im Süden des Untersuchungsraumes. So erscheint eine Grenzziehung in diesem Kontinuum schwierig. Der Autor zog die Grenze zwischen Paläotropis und Holarktis dort, wo ein Dominanzwechsel von holarktischer zu paläotropischer Flora stattfindet.

Für die Gebirge Ahaggar und Tibesti wurden eigene Analysen erstellt. Die entsprechenden Profildiagramme der Arealtypenspektren zeigen mit der Höhe einen kontinuierlichen Wandel der Flora, am deutlichsten im Tibestigebirge: von tropischen Fussstufen zu mediterran-saharischen Gipfelregionen. Auffallend ist dazu eine kontinuierliche Zunahme der Endemitenarten mit der Höhe in beiden Gebirgsstöcken. Es sind Relikte einer früheren Flora der Feuchtphasen.

Die floristischen Ergebnisse wurden zudem mit Klimaparametern vielfältig in Beziehung gesetzt, so dass auch auf die Ursachen der floristischen Differenzierung eingegangen ist.

Die Arbeit ist sowohl unter methodischen als auch unter inhaltlichen Aspekten sehr empfehlenswert für alle Geowissenschaftler, die sich mit dem saharischen Raum beschäftigen.

Dieter Klaus

Late Quaternary geomorphology and palaeoecology of the Konya basin, Turkey. By Neil Roberts. 1980. pp.296. PhD thesis, London University.
Abstract. The last two decades have witnessed several major developments in Quaternary studies, especially in low and mid-latitudes, but despite this the Near East remains a 'blank on the Pleistocene map'. The result has been not merely to hinder global palaeoclimatic reconstructions but also a failure to provide the 'environmental backcloth' crucial to our understanding of human history in this key region.

This study investigates the development of the Konya basin, which, though now dry, was formerly occupied by an extensive lake. The thesis tries to rectify a general neglect of lake basins as a source of palaeoclimatic information in the Near East, and elucidates the relationship between environmental change and the origins of agriculture in south central Anatolia. It reviews evidence of late Quaternary lake level and other palaeoenvironmental changes, and tentatively identifies a sequence of regional lacustral phases for the Near East.

The history of palaeo-lake Konya is reconstructed from fossil beaches and other shoreline features, and from cores taken through lake bottom and marginal alluvial sediments. Techniques employed include the analysis of sediments, diatoms, ostracods and mollusca, and a chronological framework is provided by radiocarbon dating and archaeological data. Overflows from a higher lake system in the Beyşehir-Suğla basin are monitored by heavy mineral analysis and by a palaeolimnological study of Lake Beyşehir.

[14]C dates on shells indicate that palaeo-lake Konya attained its last maximum level between 23 000 and 17 000 years ago. The generally cold, dry late glacial climate ameliorated after 12 000 yr BP, and water (including glacial meltwater) collected in secondary depressions to form a series of small, isolated lakes. The Konya plain has subsequently been dry, but maximum aridity apparently occurred in the early Holocene, and was followed by flooding more extensive than at the present-day during mid-Holocene times.

It is concluded that there is no direct causal link between the drying-up of the former Konya lake and the appearance of early agricultural (Neolithic) settlements on the floor of the plain. A more important factor was the creation of new land and water resources which resulted from changes in river regimes and alluvial sedimentation after the end of the Pleistocene.

Géomorphologie et palaéoécologie du bassin de Konya, Turquie, au quaternaire récent. 1980. pp.296. Thèse, Université de Londres.

Résumé. Au cours des vingt dernières années on a vu quelques développements importants dans les études quaternaires, en particulier sous les Tropiques, mais, malgré cela, le Proche Orient reste 'un vide sur la carte du Pléistocène'. La conséquence de cela n'a pas été seulement de retarder les reconstructions palaéoclimatiques mondiales, mais aussi de ne pas fournir 'la toile de fond environnementale' essentielle pour notre comprehension de l'histoire humaine de cette région importante.

Cette étude examine le développement du bassin de Konya, sec aujourd'hui mais couvert, auparavant d'un lac étendu. La thèse essaie de pallier au manque général d'études sur les bassins lacustres comme source d'information palaéoclimatique au Proche Orient, et éclaire les relations entre les variations de l'environnement et les origines de l'agriculture au centre de l'Anatolie méridionale. Elle discute l'évidence des fluctuations de niveaux des lacs au quaternaire récent et identifie provisoirement une séquence de phases lacustres pour les régions du Proche Orient. L'histoire du lac de Konya est reconstruite à l'aide de dépôts de la zone littorale et de carotte de sédiments alluviaux et bas-lacustres. Les techniques employées comprennent l'analyse des sédiments, diatomées, ostracées et mollusques, et la chronologie est fourni par les datations au radiocarbone et par les données archaéologiques. Les débordements d'un système de lacs plus élévés dans le bassin de Beyşehir-Suğla sont enregistrés par l'analyse de minéraux lourds et par une étude palaéolimnologique du lac de Beyşehir.

Les ^{14}C datations sur les coquilles indiquent que le palaéo-lac de Konya a atteint son dernier niveau maximum entre 23 000 et 17 000 ans BP. Le climat devint froid et sec par la suite, mais s'améliora après 12 000 ans BP et des eaux (y compris des eaux glaciaires) s'amassèrent dans des dépressions secondaires pour former quelques petits lacs séparés. Le bassin de Konya est sec depuis cette époque mais l'aridité la plus sévère eut lieu sans doute au cours du Holocène antérieur et des inondations importantes que celles d'aujourd'hui lui succédèrent pendant l'Holocène moyen.

Nous concluons qu'il n'y a pas de relation causale directe entre l'assèchement du lac de Konya et l'apparition des premiers villages agricoles sur le sol de la plaine. Plus important fut la création de nouvelles resources d'eau et de terre dues à des changements dans les régimes fluviaux et la sédimentation alluviale après la fin du Pléistocène.

Neil Roberts, School of Geography, Oxford, United Kingdom